CAPTAIN MALLON
DOUGHBOY HERO

by Stephen D. Chicoine

Freedom History Publishing

Published by   Freedom History Publishing
                Minneapolis, Minnesota
                freedomhistory.com

Copyright © 2017 by Stephen D. Chicoine

All rights reserved. No part of this book may be used or produced in any manner whatsoever without written permission except in the case of brief quotations embodied in critical articles and reviews.

ISBN-978-0-9993161-0-8 (paperback)

Printed in the United States of America

For My Parents,
Duane and Beatrice,
who encouraged and inspired me
to live a good life with purpose
to satisfy my intellectual curiosity
and to pursue my dreams

Contents

PREFACE 3
CH 1 - TROUBLE IN IRELAND 4
CH 2 - KANSAS, THE WILD WEST 8
CH 3 - SPANISH AMERICAN WAR 17
CH 4 - A FLURRY OF PUNCHES 26
CH 5 - MINNEAPOLIS TRADESMAN 39
CH 6 - GOING OVER THERE 44
CH 7 - BATTLE OF HAMEL 51
CH 8 - MEUSE-ARGONNE OFFENSIVE 60
CH 9 - MEDAL OF HONOR AND
ONE OF PERSHING'S 100 HEROES 73
CH 10 - WORLD WAR VETERANS 83
CH 11 - NONPARTISAN LEAGUE 105
CH 12 - RETURN TO KANSAS 136
CH 13 - DEDICATION OF VICTORY MEMORIAL DRIVE 148
CH 14 - TOMB OF THE UNKNOWN SOLDIER 155
CH 15 - HENNEPIN COUNTY COMMISSIONER 176
CH 16 - HERO OF THE DISTANT PAST 195
CH 17 - THE FINAL MONTHS 203
CH 18 - FORT SNELLING NATIONAL CEMETERY 219
CH 19 - EFFIE AND THE BOYS 228
EPILOGUE 235
POSTSCRIPT 238
SOURCES AND ACKNOWLEDGMENTS 240

PREFACE

Fog enveloped everything. Artillery shells screamed overhead. Explosions shook the earth. Flashes lit up the gray world then faded away. Whistles pierced the muted din. The doughboys rose and moved forward. Every man breathing hard. The ominous rat-ta-ta-tat of machine guns filled the air. A hailstorm of lead whizzed past. The doughboys crouched as they scrambled forward. Now and then, the thud of lead muted by flesh could be heard, followed by a moan or cry. It was impossible to know what awaited them up ahead. In the chaos, the company became scattered in small groups. Captain George Mallon pushed forward with a handful of men. His destiny awaited him. Meuse-Argonne, France, September 26, 1918.

CHAPTER ONE
TROUBLE IN IRELAND

The Mallon story begins in northernmost Ireland. Centuries of death and destruction led to anguish and sorrow as a way of life for the Irish. The English were determined to subjugate the Irish. The Irish resisted fiercely to maintain their freedom at any cost. Savagery upon savagery failed to stem the indomitable spirit of the Irish. Through it all, the Irish embraced life in the mystical beauty of their Emerald Island.

The Mallon ancestral homeland was County Tyrone, one of the six historic counties of the province of Ulaidh on the northern end of the island of Ireland. County Tyrone for centuries was the center of resistance to English dominance, occupation and colonization. They endured the raids and invasions of the fierce Vikings. The real problems for the Irish began in 1169 with the Norman invasion. From that time onward, the Irish were involved in a brutal fight to remain free and independent. The fact that they faced insurmountable odds did not dissuade them from their struggle. The Irish fought heroically and endured suffering, but would not yield. English King Henry VIII declared himself King of Ireland in 1542. Tyrone's Rebellion, which lasted from 1594 to 1603, resulted in the final defeat of Hugh O'Neill, Earl of Tyrone, and Hugh Roe O'Donnell. In 1607, Hugh O'Neill, the Earl of Tyrone, fled English-occupied Ireland with Rory O'Donnell, Hugh Roe's brother, and their families in what became known as The Flight of the Earls. King James I confiscated their vast lands and, soon after, settled large numbers of Lowland Scots in the province, which they re-named Ulster. The Scots, of Presbyterian faith through the Reformation, displaced the Irish Catholics in every aspect of the Six Counties.

In the aftermath of the Irish rebellion of 1641 and the English Civil War, the victorious Oliver Cromwell led the re-conquest of Ireland from 1649 to 1653. Countless savage atrocities occurred. After the defeat, some Irish engaged in guerrilla warfare. Cromwell's brutal suppression included the destructions of entire regions and mass executions to remove all support for the guerrillas. Famine and plague followed. On the order of forty percent of all Irish Catholics perished. Cromwell then confiscated vast tracts of remaining Irish land and restricted Irish from public office. Irish Catholics reviled Cromwell's memory thereafter.

w

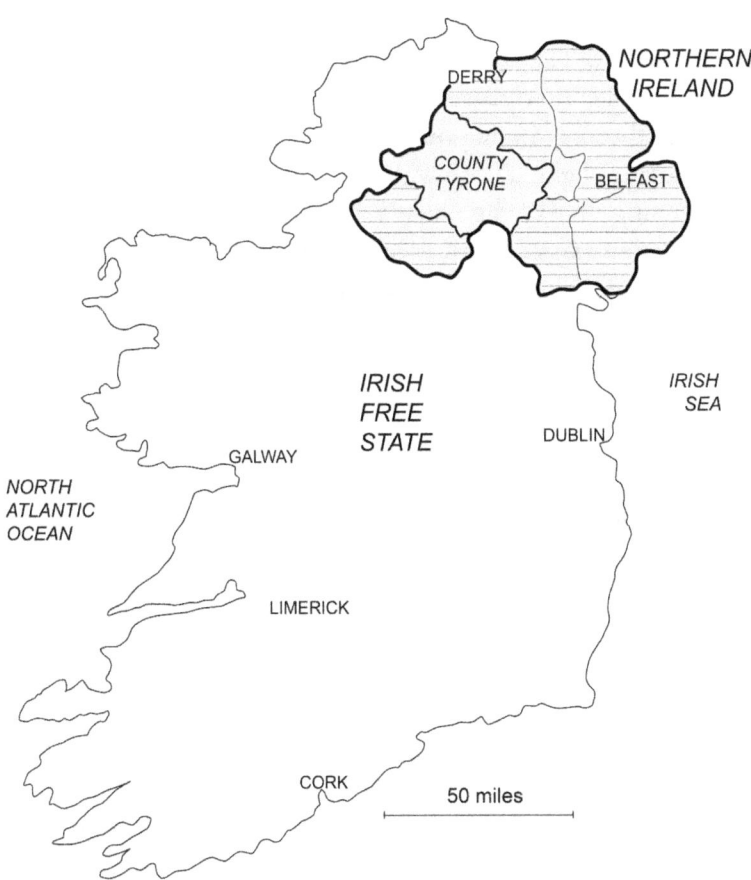

Figure 1

Henry Mallon, George H. Mallon's grandfather, was born in County Tyrone in 1792. He was a small child during the bloody Irish Rebellion of 1798 which resulted in the deaths of tens of thousands. Mallon family oral tradition tells of the harsh treatment of the children at the hands of a Protestant schoolmaster in those times. Far more ominous was the periodic appearance of marauding bands of armed Orange men invading Catholic neighborhoods to destroy and kill. The Protestant-controlled sheriffs and judges made certain the Orange men did not face prosecution for their crimes. In between raids and violence, Catholic men worked menial jobs with no hope for economic advancement. They survived just a level or two above slavery. The women did their best to keep the children from starving, but diseases claimed lives. What nourishment was available through a few potatoes disappeared with the infamous Irish potato blight. Those with the least suffer the most in such dire circumstances. The blight was widespread by 1845 and The Great Famine was underway in 1847 when Henry Mallon made the painful decision to leave his beloved Ireland and his extended family.

Robert Curry Mallon, George Mallon's father, was born in County Tyrone in 1839. He was eight years old when the family emigrated to the United States in 1847. The Mallon move was one small part of a mass response to both the oppressive conditions in Ireland and the calamity of the potato famine. Two million Irish made their way across the ocean to North America from 1846 to 1851 alone. The year in which the Mallons emigrated, 1847, was the peak year of Irish arriving in the United States. Like most Irish emigrants, the Mallons took passage on a ship to the English port of Liverpool.

At Liverpool, the Mallons and many others boarded the Kalamazoo, an ocean-going vessel bound for Philadelphia. Most Irish crossed the Atlantic in steerage, jammed below deck in unbearable conditions. The air was stale and hung heavy with the stench of human waste. Many died of disease in what the Irish referred to as "coffin ships." Mallon family oral tradition relates the story of a huge storm at sea in which the British captain lost his nerve and the Irish First Mate saved the ship from floundering and sinking.

The Mallons disembarked at Philadelphia, as did thousands of other Irish, and settled there. The city was still recovering from bloody anti-Irish riots in the summer of 1844. The Irish, many of them farmers from rural Ireland, adjusted as best they could to the crowded tenements and the filthy streets. As in the British-occupied Ireland, the Irish in the United States found themselves restricted from all but strenuous, low-paying labor jobs. The Catholic Church served as a nucleus for the Irish community. Hospitals, fire companies, benevolent societies organized around the parish church. Catholic schools protected Irish children from being forced to read the Protestant Bible in public schools and suffer at the hands of Nativist teachers.

CHAPTER TWO
KANSAS, THE WILD WEST

The boundless opportunity in the West is what made America special. In the spring of 1857, Robert Curry Mallon, who would one day become George's father, was just turning eighteen. He and his older brother John headed west to the American frontier. They drove an oxen-drawn wagon from Pennsylvania to Missouri. The Kansas-Nebraska Act of 1854 allowed settlers to decide by vote if the two territories would become slave states or free states; i.e., popular sovereignty. Ruffians of each side freely did their best to eliminate the other. Kansas Territory was known as Bleeding Kansas. Pro-slavery Missourians attacked and sacked the Freeholder town of Lawrence in 1856. The Abolitionist John Brown, in turn, followed with the Pottawatomie Massacre days later. It was into this inferno that the Mallon boys ventured when, after a year in Missouri, they headed westward into the battleground of Kansas. Pro-slavery Missourians prowled eastern Kansas and travelers made the journey with stealth and their weapons at the ready.

Pro-slavery ruffians carried multiple revolvers in their belts to give them the edge in close-quarter fighting. They would bait their target to fire off a round and then charge them and engage before the defenders could re-load. Free Soil men, when possible, armed themselves with breech-loading Sharps rifles. Abolitionists in the East shipped the rifles in crates marked "Beecher's Bibles," a reference to the abolitionist minister Henry Ward Beecher.

One hundred miles west of Lawrence, Kansas, the Smoky Hill and Republican Rivers converge to form the Kansas River. The U. S. Army established Fort Riley near that junction in 1853. The troops were to protect settlers moving westward along the Oregon Trail and the Santa Fe Trail. By the late 1850s, the troops were mainly trying to keep the violence between the pro-slavery and anti-slavery settlers from shifting westward into the region. Irish and German stonemasons and carpenters helped construct the buildings at Fort Riley and continued to work there as the fort expanded. Many those tradesmen settled on the eastern edge of the military reservation at what began the small community of Ogden, Kansas.

Robert Curry Mallon's decision, as to where in Kansas to settle, surely was based on the chaos of the time. They traveled westward beyond the battleground that was eastern Kansas and continued past Topeka and Manhattan. Robert and his brother John settled along Seven Mile Creek, north of the new settlement of Ogden, Kansas. Less than a mile to the west was the military post of Fort Riley.

For the most part, Free Soil men predominated the region, although Ogden residents were said to be more pro-slavery. In any case, the Mallon boys settled two and a half miles north of town on the creek along the eastern border of the military reservation. They raised wheat, as did most farmers in Kansas. Robert and John built stone houses from outcrops of available limestone along the creek. George's dad, Robert, had experience working stone. Seven Mile Creek was an advantageous location. The U.S. Cavalry was relatively close by in the event that Free Soil or pro-slavery ruffians appeared, as well as Kansan, Kiowa and Cheyenne raiding parties.

Jesuit priests from St. Mary's, fifty miles to the east, began to travel regularly to Fort Riley within a year of the fort being established. Their responsibility covered fourteen counties. The priest also would stop at Ogden to say Mass. In 1854, a priest consecrated the ground that became St. Patrick's Cemetery north of Ogden and just south of the Mallon farms.

The men of Ogden organized a company of volunteers when the Civil War broke out. They became the core of Company G, 10th Kansas Infantry Regiment. The 10th Kansas saw action at Prairie Grove, skirmished with Quantrill and then fought at the Battles of Franklin and Nashville, ending the war in the Mobile campaign. The Mallons chose to not volunteer for military service. Conscription applied only to citizens and to immigrants, who had applied for citizenship.

Soon after the Civil War's end, more Mallons moved west to Kansas. Robert Curry Mallon's sister Jane Elizabeth and his cousin Patrick Henry Mallon, both also born in Ireland, settled on Seven Mile Creek. The old Irish patriarch Henry Mallon and his wife Margaret Mary left Philadelphia and settled in Kansas by June 1865. Henry was an American citizen, having naturalized in 1856 in Philadelphia. In 1866, Father Louis Dumortier, S.J., arranged the construction of a 30-foot by 16-foot stone building in Ogden, which he consecrated St. Patrick's Church. The Catholic community consisted of twenty-seven families, several of which were Mallons.

Jane Elizabeth Mallon's husband, Felix Boller, ran a butcher shop in Ogden. He also served as town marshal. There was a brewery in town, which could only spell trouble with so many soldiers nearby at Fort Riley. In December 1867, some drunken soldiers and their officer were "harassing" a woman in her home. Marshal Boller responded to a call for help. The soldiers shot and killed Felix Boller, leaving Jane Mallon Boller a widow with four small children. She traveled to Washington for restitution, but received nothing. Mallon family oral tradition recalls that an enraged Jane tore an American flag from its standard in the halls of the U.S. Congress as she left. The family buried Felix in St. Patrick's Cemetery.

Robert Curry Mallon, a single man, continued to farm with his older brother John Mallon. Their father Henry Mallon died on April 18, 1873. The family laid Henry to rest in St. Patrick's Cemetery. Just a few months after his father's passing, Robert Curry Mallon married Emma Stephens of Junction City, Kansas on August 31, 1873. Robert was thirty-four years old and Emma was twenty. She also was non-Catholic. Emma's parents, George Stephens and Mary Smith, were both born in Ireland. George was a Scot, whose ancestors settled in occupied Ulster. Henry Mallon likely would have struggled with accepting the marriage, had he still been alive. Family oral tradition relates that

George Stephens was strongly anti-Catholic.

Emma's father, George Stephens, served in the Union Army during the Civil War. James Cameron organized the 79th New York Cameron Highlanders, enlisting Scots settled in New York City. They wore Tartan trousers and the traditional Glengarry cap and marched to the tune of bagpipes. The 79th New York fought at First Manassas in July 1861 when the Confederates routed the Union Army. The regiment went south down the coast as part of Sherman's expedition to Port Royal, South Carolina. On September 1, 1862, in the aftermath of the Confederate victory at Second Manassas, the 79th fought desperately at Chantilly and succeeded in preventing Stonewall Jackson from cutting off the retreat of the Union Army. The 79th suffered great loss in the close quarters fighting. In that battle, Private Stephens was taken prisoner. He came home after a prisoner exchange and was discharged for disability in February 1863.

Robert and Emma's first child, a daughter Ida, was born on Seven Mile Creek in 1874. George Henry Mallon, their second son, was born in 1877. He grew up on the farm, working hard and growing strong. The Robert Curry Mallons lived in a six-room stone house, which Robert built with his own hands. The nomadic tribes and bison were gone from the eastern edge of the Great Plains, but the American West remained and the Mallons flourished in the setting. George Henry Mallon grew up on the family farm on Seven Mile Creek among many cousins. Besides his Uncle John Mallon's family, there was the family of their cousin Patrick Mallon. George's first cousin Margaret became "Liz Williams", a champion quarter horse rider for a Wild West show. His first cousin John, known as "Topeka Johnny", served as a peace officer in Indian Territory under Deputy U.S. Marshal Bill Tilghman.

Nearby Fort Riley continued to play an important role as a military post after the Civil War. Lt. General Philip Sheridan recommended to Congress in 1884 that Fort Riley be made the cavalry headquarters of the United States Army. That solidified the future of the now venerable army post.

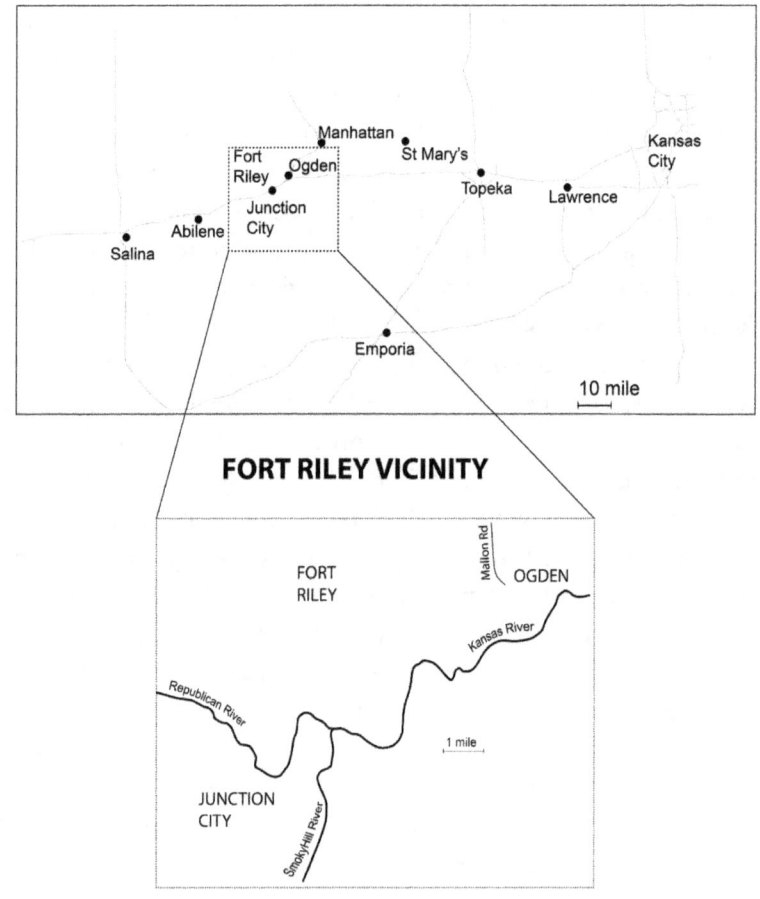

Figure 2

Junction City on the southern edge of Fort Riley was ten miles from the Mallon farms. The boom town fielded a baseball team as early as 1867. They played at Athletic Field as part of the Kansas State League against teams from Manhattan, Topeka, Abilene, Emporia and Leavenworth. The Abilene, Kansas baseball team traveled to Junction City for a game as part of the festivities for the Fourth of July in 1891. Abilene won a "hotly contested game." Other events included a footrace and a tug-of-war.

Junction City also regularly played the Fort Riley baseball team. By the mid 1890s, when George Mallon was in his late teens, baseball was on its way to becoming the national pastime. Baseball games were particularly major social events in rural America. In May 1895, Junction City threw a farewell part for the 7th Cavalry. All the troopers attended a huge city-wide picnic and enjoyed a baseball game between the Junction City nine and the Fort Riley team.

The *Topeka State Journal* called a game played in 1896 between Junction City and Emporia as "one of the finest ever played in the city." Junction City won. Kansans followed the St. Louis Browns in the major league National League. Every team included Irish Americans on its roster. The Baltimore Orioles won their third straight league pennant in 1896. Two of their stars later joined the Baseball Hall of Fame – John McGraw and Wee Willie Keeler (his name Americanized from O'Kelleher). Pitcher Joe Corbett, younger brother of boxer James J. Corbett, also played on the Orioles. The leading power hitter in the league was Big Ed Delahanty of the Philadelphia Phillies. He also was inducted into the Baseball Hall of Fame.

The location of the Mallon farm on the eastern edge of Fort Riley Military Reservation surely impacted young George Mallon in many ways. Among the local legends, whom George likely came to know, was Conrad Schmidt. He was an old retired cavalry sergeant, who by 1889 worked as range rider for Fort Riley. George Mallon was twelve years old when Schmidt began riding the perimeter of the vast military reservation to keep out trespassers, including cattle and poachers. Sergeant Schmidt, wearing a floppy broad rim hat and ambling along on his mule, passed along the Mallon farm on a regular basis. The old cavalryman, a Civil War veteran, circumvented Fort Riley for nearly twenty years.

Conrad Schmidt first drove an ox-drawn wagon as part of the 1857 expedition against the Mormons in Utah. Just months before the Civil War broke out, Schmidt enlisted in the U.S. Army at Fort Leavenworth, Kansas on February 5, 1861 as a trooper in the Second Dragoons. The unit was re-designated the 2nd U.S. Cavalry Regiment in August. Schmidt and his comrades saw action in the Peninsula Campaign, at Second Bull Run and Antietam. The regiment performed admirably in the major cavalry battle of Brandy Station. Schmidt scouted at Gettysburg in 1863 and was responsible for locating Jeb Stuart's rebel cavalry, contributing to the Union victory. He was wounded in the major cavalry battle at Trivelian Station in June 1864. He distinguished himself at the Battle of Third Winchester on September 19, 1864. By 1864, Schmidt was first sergeant. While on scout four days later, he and three others were captured by Mosby's Rangers. The guerillas murdered the other three. The bullet, intended for Schmidt's forehead, entered through his upper lip, ripped out his teeth and exited through his ear. The horses trampled his body on the ground, but he survived. First Sergeant Schmidt recovered and returned to the regiment to serve out his enlistment. The chloride of lime, which the doctors used as a disinfectant, destroyed what remained of Schmidt's teeth. He could only eat soft food for the rest of his life and mainly subsisted on liquids. He lost his injured kidney later in life.

Conrad Schmidt was discharged from the U.S. Army in early 1866. A few months later, in July, he married 19-year-old Mary Rush at Junction City, Kansas, just south of Fort Riley. The couple homesteaded south of Ogden on Humboldt Creek. He and Mary lost two infants in 1869 and in 1875, both of whom they buried in St. Patrick's Cemetery, just south of the Mallon homesteads. Mary passed away in 1882 at the age of 35, leaving

Conrad to raise six children with ages from five to eighteen years. He worked as chief clerk for the Quartermaster at Fort Riley for a time. Conrad retired in 1889 at age 60. In retirement, he went to work as a range rider for the Fort Riley Military Reservation.

A local newspaper referred to "Conrad's pleasant face, accompanied by the old-time sweet smile for the ladies." Exchanges with the farmers would have provided a break in the monotony for both the range rider and the farmers. George's mother and grandmother might have welcomed a rider with a cool glass of water and perhaps a freshly baked cake. Conrad Schmidt, no doubt, came to know George Mallon's grandfather, George Stephens, himself a Civil War veteran. Both Schmidt and Stephens were member of the Union Army veterans group, the Grand Army of the Republic (GAR).

History was not yet ready to allow Schmidt to fade into obscurity as a range rider. In 1890, decorated army officers formed The Medal of Honor Legion. That led to a surge in nominations from aging Union Civil War veterans for the War Department to review. In 1894, the War Department awarded the Medal of Honor to Colonel (Retired) Theophilus Rodenbough for his bravery at the Trivelian Station, thirty years earlier, on June 12, 1864. Rodenbough was a captain at the time, commanding the 2nd U.S. Cavalry. Two years later, in 1896, Rodenbough and two other officers of the 2nd U.S. Cavalry, including Captain Robert S. Smith, offered personal testimonies as to Sergeant Conrad Schmidt's valor at Third Winchester. Captain Rodenbough, just recovered from his wounds at Trivelian Station, led his regiment in a great charge against the Confederate position in the Third Battle of Winchester on September 19, 1864. A bullet ripped through Rodenbourgh's arm and another killed his horse. First Sergeant Schmidt rode into a hailstorm of lead to within just a few yards of the enemy position, pulled the captain up onto his horse and rode off to safety.

For this, Conrad Schmidt received the Medal of Honor in 1896. Thirty-two years after his gallantry, Conrad Schmidt became a local celebrity in the Fort Riley-Junction City-Ogden area.

George Mallon was nineteen years old in 1896. Conrad Schmidt had been riding the range around Fort Riley's perimeter since 1887. Mallon and Schmidt would have known each other. The reverence and deference by all for a Medal of Honor recipient, as well as the admiration, which George Mallon himself felt for the old sergeant, would not have been lost on the nineteen-year-old. He yearned to leave the farm and find adventure.

George Mallon most certainly also heard war stories from his Grandfather George Stephens. Those stories should have dispelled any myth of the glory of war, given the bloodshed and bitter defeats which Stephens experienced at First and Second Manassas, including being taken prisoner. A newspaper man, who later interviewed Mallon, wrote of him, "One of the prized mementoes that he carries with him is the little G.A.R. button that his grandfather wore." A medallion of the Grand Army of the Republic, the organization of Union Army veterans, stands in the ground next to the gravestone of George Stephens. The Stephens, George's maternal grandparents, being Protestant, were buried in Ogden Cemetery.

Sergeant Conrad Schmidt died in 1908 and was buried in St. Patrick's Cemetery along Seven Mile Creek. The remains of over forty Mallons lie in Saint Patrick's Cemetery, which dates from 1854. Among those are George Mallon's paternal grandparents, whom George never knew, and George's parents, Robert Curry Mallon, who died in 1914, and Emma, who died in 1935.

Mallon Road remains to this day north of Ogden, Kansas, marking the land settled in the early days by George H. Mallon's father, Robert Curry Mallon.

CHAPTER THREE
SPANISH AMERICAN WAR

George Mallon was twenty-one years old in 1898. He was a farm boy, five-feet-eight-inches tall and well-built with broad shoulders and a thick body forged on the farm. He imagined a life beyond the wheat farm. His Grandfather Stephens of the 79th New York Infantry was seventy-nine years old and still in George Mallon's life. Conrad Schmidt, the old Civil War cavalry trooper, was sixty-eight years old.

Anyone in the United States, who read the newspaper, was aware of the outrages, which the Spanish were committing in Cuba. The Cubans revolted in 1895 after years of atrocities. Spanish reprisals included executions and concentration camps. In January 1898, the United States government sent the battleship USS Maine to Havana harbor, supposedly to protect American interests. How exactly was unclear. On February 15, 1898, the USS Maine exploded violently, killing 268 American sailors. After naval boards of inquiry, considerable saber rattling in the newspapers and much debate, the United States Congress declared war on Spain on April 25, 1898. In the emotional aftermath of the Maine disaster, men rushed to serve. They wanted to avenge the Maine and serve their nation. They also felt the call in the name of freedom and justice to assist the helpless Cuban patriots.

George Mallon enlisted and mustered into the 22nd Kansas Volunteer Infantry Regiment on June 18, 1898. The Spanish American War was yet to fully get underway. Commodore George Dewey won the Naval Battle of Manila Bay in the Philippines on May 1. The U.S. Army had not yet landed in Cuba when Private Mallon mustered into the military service. The landing took place on June 22, just four days after Mallon's enlistment. Events moved quickly. The decisive Battle of San

Juan Hill took place on July 1. Spain and the United States agreed to end hostilities on August 12.

Private Mallon and the 22nd Kansas Infantry never made it to Cuba. They moved to Camp Alger near Falls Church, Virginia. There was insufficient water supply on the site. The weather was hot and humid. Flies were everywhere. Sanitation facilities were inadequate and a typhoid epidemic broke out. Fourteen men of the 22nd Kansas died. The entire corps moved at the beginning of August to Camp Meade, south of Harrisburg, Pennsylvania. Spain agreed to end the war on the day after their arrival at Camp Meade. In any case, conditions at Camp Meade were much better in every sense, as they should have been. On September 9, the regiment left Camp Meade and returned to Kansas. They mustered out at Fort Leavenworth on November 3, 1898. George returned home to meet his newest sibling, born in March, the tenth child of Robert Curry Mallon and Emma Stephens.

George Mallon was determined that the 22nd Kansas Volunteers experience would not define his military service to his nation. The 12th U.S. Infantry was heading to the Philippine Islands, where the Filipinos were resisting the American occupation. Mallon enlisted at Fort Riley on January 7, 1899, just two months after being discharged from the Volunteers. He was assigned to Company K. Among other enlistees was Private Walter Krueger in Company M. Krueger would rise to command the Sixth Army in the Philippines during the Second World War and retire as a full general in 1946. On March 12, 1899, the 12th U.S. Infantry Regiment left New York, bound for the Philippines. They passed through the Suez Canal, a noteworthy event for anyone at the turn of the century. Thirty-two long days after their departure, the men of the 12th arrived at Manila on April 14. It was the hottest time of year in the Philippines.

Sergeant George H. Mallon, 12th U.S. Regt.
Courtesy of Diane Gossage

Major General Arthur MacArthur's Malolos Campaign was underway. Among the units involved was the 20th Kansas Volunteers. The Americans took the Filipino capital of Malolos two weeks before the 12th Infantry landed. The Filipino freedom fighters escaped the American attempt to seal off and destroy them. Mallon and his comrades entered the field soon thereafter and moved north from Manila through San Fernando and San Isidro.

The regiment took part in numerous actions against guerillas across the island of Luzon. They forded rivers and endured the rainy season with only the protection of their slouch hats and ponchos. The foremost military objective on Luzon was the re-located capital of the First Philippine Republic in Angeles, fifty miles north of Manila. In mid-August, the 12th Infantry reached Angeles. This was to be George Mallon's real baptism of fire. He and his comrades advanced across open rice fields against 2,500 entrenched Filipino freedom fighters. Skirmishing began early in the morning. The main assault commenced at 1030 hours. The regulars coolly delivered withering volleys of rifle fire into the defenders while American artillery pounded the Filipino position. The Filipinos finally withdrew and the 12th Infantry took the city, ending the five-month campaign. George Mallon was a combat veteran.

The monsoon rains made any further advance possible. Fighting continued while the Americans held Angeles. Guerrillas probed and harassed the Americans during the typhoon season through September and October. On August 19, Lt. Alfred W. Drew, commanding Company I, 12th Infantry, was killed in action outside of Angeles.

A soldier does as he is told, regardless of his opinion of the merits or righteousness of the campaign or war. This is particularly true in the case of regulars. In a democracy, citizens have a right to know and a duty to express their opinions. Many in the United States questioned the war in the Philippines. They recognized the conflict as the War for Philippine Independence. They saw the war for what it was; i.e., a war of conquest, the first beginnings of empire. Concerned citizens formed The American Anti-Imperialist League. Among the more distinguished members were Mark Twain, former President Grover Cleveland, steel magnate Andrew Carnegie and labor leader Samuel Gompers. Twain wrote:

> ... here are a people who have suffered for three centuries. We can make them as free as ourselves, give them a government and country of their own, put a miniature constitution of their own, put a miniature of the American constitution afloat in the Pacific, start a brand-new republic to take its place among the free nations of the world.

The League published a volume in 1899 with letters sent home by men of the 20th Kansas Volunteers and the 13th Minnesota Volunteers. The soldiers lamented the moral position in which they found themselves, killing those fighting for their freedom and independence.

Meanwhile, George Mallon and the regulars of the 12th U.S. Infantry Regiment prepared to resume the campaign. In early November 1899, the 12th Infantry pushed out from Angeles to begin the Tarlac Campaign. Bull carts – wooden carts pulled by carabao or water buffalo – served as the regimental supply train. A late typhoon season resulted in rivers rising dramatically and washing out roads. A soldier of the 12th Infantry wrote home:

> ... we have been on the move since November 10th ... The next morning early, we started out to take Bamban where they were strongly intrenched on the opposite side of a deep and swift running mountain stream. Here under a heavy fire from our own artillery they were flanked and driven out ... it was a hot old time around there for about three hours. That was the last scrapping that was done by this regiment.

After the assault on Bamban, the regiment labored in the searing heat to build a corduroy road (log planks perpendicular to the direction of the road) across an otherwise impassable swamp. "If the situation had not been so dangerous, it would have been exceedingly ludicrous …". The regiment, having accomplished their goal, slept that night and arose early morning to make a forced march of fifteen miles. Their objective was Tarlac, the relocated capital of the insurgent leader Emilio Aguinaldo. There was great expectation of another grim assault. Instead, they found that Aguinaldo and his men slipped off into the jungle to fight another day. The regiment made a dramatic show of force, marching into Tarlac four abreast. The campaign ended on November 20 with all objectives secured.

After a week in Tarlac, the regiment marched south to Panique near Angeles. The engineers had restored the rail line so the 12th Infantry rode the train north from Panique to Bayambang. The Agno River was pristine and the men enjoyed the break.

The 25th U.S. Infantry (Colored) also was on Luzon. Before leaving Manila for the field, the regimental baseball team played several games. "We defeated all comers until orders came for the regiment to advance." The regiment then fought their way north from Bamban to Dagupan. The 25th ended up at Bayambang at the same time as the 12th U.S. Infantry. The 25th historian wrote: "Soon after our arrival, a game was arranged with our old rival and best friend, the 12th Infantry." The regiments knew each other well, the 12th having bested the 25th in two baseball games at Camp Chickamauga in Georgia in 1898 before their landing in Cuba. All agreed that they meet again on the diamond. "The game was played with that fighting spirit that both regiments are noted for." The 25th won the game, resulting in the 12th insisting in a re-match on Christmas Day. Again the 25th triumphed. The regiment claimed the baseball championship in the Philippines for 1899, 1900 and 1901.

After the game, the 12th Infantry regulars sat down to a

Christmas meal of stewed corn, macaroni and cheese, fricassee chicken, eggs to order, roast beef, garden peas and, of course, coffee and cigars. There is no mention in the historical record as to whether the regulars of the 25th Regiment joined the 12th Regiment in the meal. Nor is it known whether the powerfully-built George Mallon played in the baseball games. Certainly, hundreds of young men in rural Kansas grew up playing baseball in the late 1800s.

The U.S. military recognized the importance of athletic competition for the many young men assigned to lengthy postings in the Philippines. Sports not only encouraged camaraderie and boosted spirits, it also was intended as a possible alternative to drinking, womanizing (and sexually transmitted diseases), desertion and suicide. Muscular Christianity was an important force in America from 1880 until 1920. President Teddy Roosevelt's father, Theodore Sr., known as Thee, was an ardent disciple of Muscular Christianity and used his considerable wealth and influence to promote the movement. As a result, the president was devoted to physical fitness and manliness. He and others believed that the re-discovery of masculinity through athletic competition could offset the negative influences of modern society and restore America's moral fiber.

Sometime between late 1899 to late 1901, George Mallon won his Army boxing championship in the Philippines. A 1920 newspaper article about George H. Mallon read: "Among other things, George cherishes a belt to show that he was boxing champion of the Philippine Islands." The details are lost to history. It is documented that African Americans regular Army soldiers brought boxing gloves with them to the Philippines.

Theodore Roosevelt, who assumed the presidency in September 1901, remarked of boxing:

> I thoroughly believe in boxing, exactly as I believe in football and other rough, manly games, and I think it the greatest mistake that decent people should ever allow the hard-hitting, game qualities

which make a man a man, to be monopolized by men who don't believe in decency.

The 12th and the 25th were on patrol constantly in early 1900. There were countless ambushes and firefights in the bush. In late 1900, the 1st Battalion of the 12th U.S. Infantry went to the Island of Samar to crush resistance in response to the massacre of a company of the 9th U.S. Infantry regiment at Balangiga on September 28, 1901. Sergeant Mallon, whose company was in 2nd Battalion, was fortunate. The 2nd Battalion did not go to Samar. The resulting campaign was one of barbaric reprisals.

In October 1900, Twain published an op-ed in *The New York Times* in which he lamented the Treaty of Paris. "… I have seen that we do not intend to free, but to subjugate, the people of the Philippines. We have gone there to conquer, not to redeem … so I am an anti-imperialist." After Admiral George Dewey withdrew, William Jennings Bryan became the Democratic candidate for the November 1900 presidential election. He made anti-imperialism a key part of his platform. The incumbent Republican president, William McKinley, with New York Governor Theodore Roosevelt as his running mate, won the election. While there were other issues, such as the silver versus the gold standard, the election was considered a mandate for imperialism and the Anti-Imperial League began a long, slow decline.

Through all the intense political debate back home in the States, George Mallon was still in the fight in the Philippines. He left no known statement as to what he thought about the American role to suppress the Philippine's independence movement. He was well aware that many of the "uncivilized" Filipinos were Catholics, as was he. And Mallon certainly understood anti-Catholic injustice from his Gaelic father.

In late 1900, a bullet struck Sergeant George Mallon in the chest during a firefight. He carried that bullet in his chest for the rest of his life. Sergeant Mallon received his discharge from the service on January 6, 1902 while on a ship heading home from the Philippines. He was listed as an invalid. Mallon received special mention for bravery in the many engagements in which he participated.

The belt from Mallon's boxing success in the Philippines served as a reminder to him of his prowess and success in the ring. That, no doubt, encouraged him to consider pursuing the sport further once he returned home.

CHAPTER FOUR
A FLURRY OF PUNCHES

George Mallon had no intention of returning to Kansas to farm upon his return from the Philippines in 1902. Perhaps George had seen and experienced too much. Perhaps he had dreams. As the oldest of ten children, he moved on to find his future. He headed for the big city.

The Philippines experience was a turning point in George Mallon's young life. He saw his first combat and he learned that his stocky build and brawling style lent itself to the sport of boxing. It is likely that he also discovered beer. Manila was the site of San Miguel brewery, founded in 1890. George's home state of Kansas had been dry since 1881 when George was four years old. It was the first state to adopt Prohibition. Among the many establishments shut down was the brewery in Ogden, which early settler Theodore Weichselbaum established.

By 1903, Mallon was living in St. Louis, Missouri, where he worked for the Transit Company as a motorman. St. Louis was a city of breweries, the acknowledged world's largest producer of beer. The predominant producer was the Anheuser-Busch Brewing Association. A major competitor in town was Lemp Brewery, the predecessor of Falstaff Brewing Company. St. Louis also had a well-earned reputation as a workingman's town. Beer and unions were its icons.

Is there no such thing as bad press? The front-page coverage, which George Mallon received in the March 25, 1903 issue of the *St. Louis Republic,* would have enhanced his reputation as a brawler. The headline on the article read:

POLICE HAVE BATTLE WITH TWO MOTORMEN
Trouble Following Incident at Broadway and Locust is Resumed
at Car Barns
PISTOLS AND FISTS USED

Patrolman Frederick Luke was interviewing George Schwartz, who claimed he was knocked from his wagon. While the policeman was jotting down some notes, Motorman John H. Burns, obviously agitated, snatched the notebook. Patrolman Luke and Motorman Burns "fell together in a heap upon the sidewalk." A large crowd witnessed the scuffle. The policeman recovered his notebook and told the motorman to take his streetcar back to the barn. He went with Burns to the barn and, once there, attempted to place him under arrest. Burns resisted and the two became involved in a second fight. The *Republic* read: "During the scuffle, George H. Mallon, another motorman, rushed to the rescue of Burns. According to the police report, he drew a pistol from the inside of his coat pocket." Patrolman Luke knocked the pistol from Mallon's hand and went after Mallon. The two went after each other and traded punches. Luke yelled out for help. A beat cop heard the call and rushed into the barn. "The two battled with the men for ten minutes before they could subdue them." The police locked up Burns and Mallon at the Eighth District Police Station. They charged Mallon with carrying a concealed weapon and Burns with disturbing the peace and resisting an officer. The newspaper listed George H. Mallon's address, 3129 Laclede Avenue. What transpired legally, as a consequence of the brawl, is unknown.

St. Louis was the site of the 1904 Louisiana Purchase Exposition, a world's fair to commemorate the 100th anniversary in 1903 of the acquisition of the central core of the United States. George Mallon went to work for the St. Louis police force during the 1904 World Fair. The World's Fair in St. Louis was a grandiose celebration of the new American empire in the aftermath of the Spanish American War. The fair included natives and displays from the newly-acquired territories of Guam and Puerto Rico, as well as the Philippines. In fact, the Filipino struggle for independence continued, despite President Theodore Roosevelt's declaration of July 2, 1902 that the war was officially over.

The exposition's vast complex included 1,500 buildings, many of them lit up by the new wonder of Thomas Edison's electricity. Every state in the union showcased its finest. Twenty nations had pavilions. The largest of all the exhibits was that of the Philippines. The 47-acre site with its own lake was home for seven months to 1,100 Filipinos. The experience must have been most interesting to a recently returned combat veteran, such as George Mallon, who trudged for weeks through the jungles. No doubt, the exhibit bore little similarity to the reality.

Numerous native fishing boats bobbed on the water. The center of the exhibit was a replica of Manila with a reproduction of the cathedral and the government offices. The Philippines Scouts and the Constabulary, 700 in number, drilled in precision formation in their starched khaki uniforms waving the American flag to the delight of viewers. They fought alongside the U.S. Army against the insurgents to suppress the independence movement. Six Philippine villages surrounded Manila. Eighty Visayans made up on village and eighty Moros another. What really drew the crowd was the village of the nearly-naked Igorots dancing to their drums. The real draw was that the head-hunting ate dogs. This was normally only for ceremonial purposes, but the promoters had the Igorots butcher, cook and eat dogs every day for the enjoyment of the fairgoers. The Bagobos village featured tall warriors with long hair and elaborately beaded clothing. Then there was the Moro village. No one mentioned that the Moro Rebellion continued in the Philippines.

Another feature of the St. Louis Fair was the aging Chiricahua Apache leader Geronimo, seventy-five years old at the time. While the Philippine natives were essentially spoils of war, Geronimo remained officially a prisoner of war. The government kept him at Fort Sill and refused to ever let him return to his homeland in the Southwest. Like the Filipinos, Geronimo was a war trophy. Perhaps the appearance, which most competed with the dog-eating Igorots, was the airplane. The Wright Brothers made their famous flight at Kitty Hawk on December 17, 1903. So only six months later, Americans, including George Mallon, were treated to seeing man in flight over the fairgrounds. The St. Louis fair popularized the hot dog, French's mustard, cotton candy, the ice cream cone, Dr. Pepper and puffed rice cereal. And, of course, there was beer at every turn. All this and more became part of George Mallon's experience as a newly-returned combat veteran.

An article back on page 14 of the October 8, 1904 issue of the *St. Louis Republic* reported that, among others, Emergency Special George H. Mallon of Second District was charged with intoxication, fined $25 and reprimanded. After the fair's closing in December 1904, Mallon returned to his job as a motorman on the streetcars. That was simply a job to put food on the table. He wanted more.

The smell of sweat and blood, the dim lighting from weak bulbs in cages high on the ceiling, the echoing sounds of grunts from exertion, of weights clanking and of leather gloves hitting leather bags. It was in a gymnasium in St. Louis, where George Mallon found renewed passion for boxing. A professional fighter noticed Mallon working on a bag and convinced him to put on the gloves and get into the ring with him. They sparred a few rounds. The fighter liked Mallon's powerful punches. A friendship developed. Irish-American Jack Dunleavy began to coach Mallon. He taught him the subtleties of the left jab and left hook, some basic footwork and blocking.

An interview with Jack Dunleavy in The *St. Louis Republic*, from several years before he met Mallon, offered insight into his fight style. The headline read: "Jack Dunleavy, Fighter, Believes that Right Hook and Left Hooks and Jabs Are Among the Most Important Blows of the Day."

> ... the most important blow of recent battles have been short-arm hooks and jabs, which start but a few inches from the opponent's jaw or body. In nearly every recent fight of importance, he claims, these blows have played the leading part ... draws but his arm but a trifle before delivering the blow, which was a short-arm affair... It was a short blow, but it did the business ... a short right hook with the force of the body and shoulders behind it ...

> The hook and short-arm jab, used without drawing the fist back at all, but giving it force by using the shoulders and weight of his body, is the most deadly blow in the ring today. It is easier to land than a swing ... a steady succession of hooks in close work before the swing was landed.

The power of a hook comes from holding the arm at ninety-degrees and pivoting on the lead foot, thereby rotating one's body into the punch. Dunleavy believed that Mallon's thick body and his hammer fists would allow him to be particularly adept at destroying opponents with a hook. While the hook is a difficult punch to throw, a skilled boxer like Jack Dunleavy could work on Mallon's technique. However, it was George Mallon's upper cut, which was the key to his success.

It was Dunleavy, who convinced George Mallon to enter the St. Louis amateur boxing tournament. George Mallon's first heavyweight fight was with Harry Chalmers. Chalmers knocked Mallon to the ground at the beginning of the first round. Mallon got back up and landed a strong right upper cut to Chalmers' jaw. Chalmers fell to the ground, unconscious. The fight was over with time remaining in the first round. Mallon's next opponent was Joe Ledden, a well-regarded amateur fighter. Ledden apparently taunted Mallon, telling him that he was going to knock him out. In the first minute of the first round, Mallon got in close and landed his upper-cut to Ledden's jaw. His powerful blow lifted Ledden a foot off the ground. Mallon similarly dispatched a tough square-jawed heavyweight fighter named Michaels. A sportswriter wrote that Mallon's performance was among the best he had ever seen in an amateur ring.

> Mallon is a well-formed man ... He carries his right arm against the side of his chest, the fist almost in line with the front of his body. He gets in close to the man and then shoots it out from that position.

The March 18, 1905 issue of the *St. Louis Republic* mentioned "George Mallon, the champion amateur ..." sparring for three rounds with Jack Dunleavy. The main event on the card that night at Tivoli Hall was between a local wrestler George Baptiste and Arata Suzuki, a Jiu-Jitsu martial artist from Japan. To the great joy of the Americans, the wrestler won.

The Leavenworth (Kansas) Post of July 3, 1906 reported, "A good many from Leavenworth will attend the boxing matches in Praun's Grove near Kansas City ... ". Praun's Grove was out beyond the end of the Chelsea Park streetcar line out beyond the edge of the metro. The papers touted George Mallon as the St. Louis fighter, the Kansas City, Kansas fighter or an Ogden boy. He faced Spike Kennedy, a well-regarded fighter with a national presence, in a welterweight match.

The news media set up the fight with glowing praise for Mallon:

MALLON-KENNEDY
FIGHT AT PRAUN'S GROVE TOMORROW
SHOULD BE A GOOD MILL
MALLON IS IN FINE CONDITION – SAYS
HE WILL SURELY WIN

George Mallon of St. Louis, who is in training at the fort gymnasium for his fight with "Spike" Kennedy of Kansas City, Mo. at Praun's Grove on the K.C.-W. line this side of Kansas City, Kan., finished his work yesterday and will go into the ring at 3 o'clock tomorrow afternoon in the finest kind of shape. He, Billy Rhodes, Jack Dunleavy, Jack Purtle and Young Sharkey have been working out together at the Fort and the other fighters say if Kennedy puts Mallon out, they won't let Mallon come back again, but George says he'll be back, so therefore he intends winning.

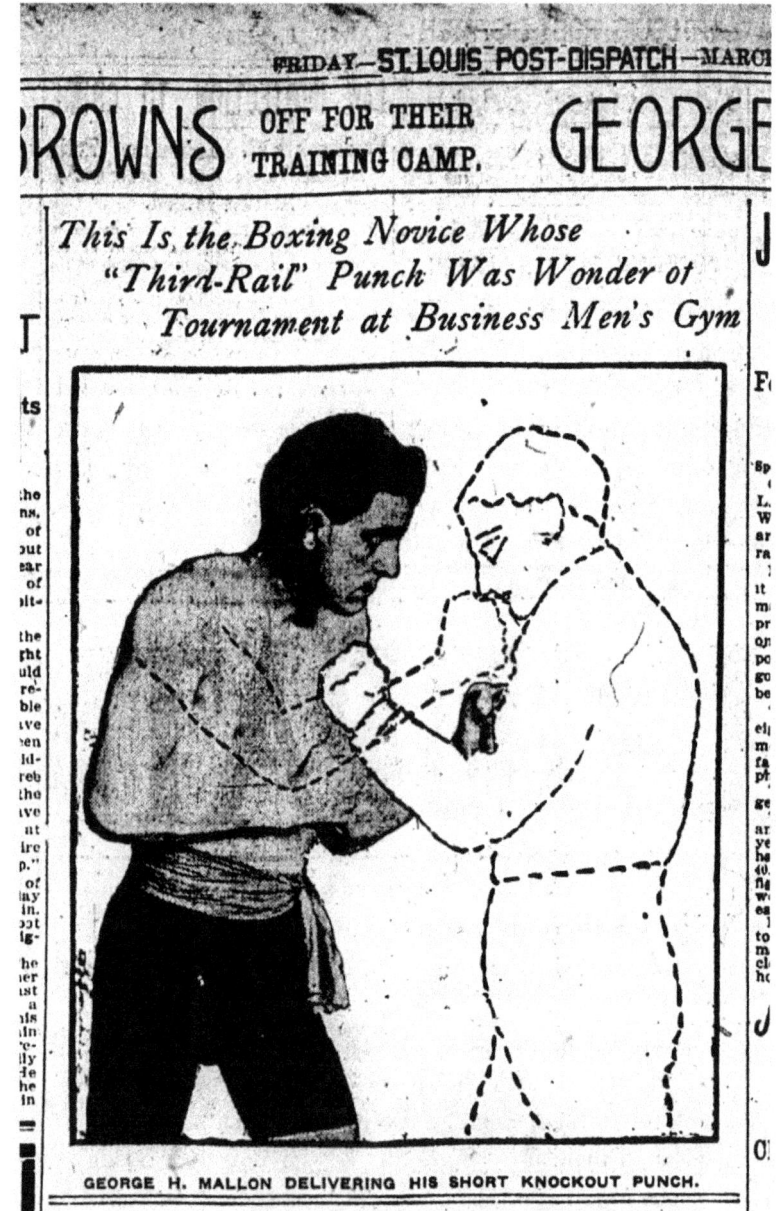

Sketch of George H. Mallon's Famous Knockout Punch from the *St. Louis Post-Dispatch*, March 3, 1905, pg. 7

Mallon is a finely featured young man of good height and a tremendous amount of steam back of his blows. Billy Rhodes says he thinks Mallon could drive a steam engine up a track a mile if he jolted one on the tender. Mallon has made many friends during his short stay here and tickets for the fight on sale in several saloons here are going fast. Quite a bunch of local sports will go down to see the mill.

The Mallon – Kennedy fight, the main event of the night, was billed as 20-round match. There were some well-known boxing figures in the crowd. An estimated $1,800 in bets were placed before the fight began. The odds were ten to seven. The fight began at 4:15 with a flurry of punches. Mallon was confident and aggressive. The brawler waded in and soon landed some solid blows on Kennedy. Spike's strategy seemed to be to block at first to see how Mallon threw his punches. There was no slipping, no counter punching. Kennedy blocked Mallon's punches or did his best to roll with them. He was an experienced fighter and was careful to not leave himself open to Mallon's devastating upper cut. Most observers felt that Mallon might have won the first round, but the bell ended the round with no single defining blow. The second round of the bout began in much the same way. One reporter wrote, "Mallon started out at a furious pace and had a shade the better of the first round and half of the second...". Kennedy began to parry Mallon's punches and throw counter punches. Mallon did not flinch, but the momentum began to shift. A couple of Kennedy's punches landed with effect. Mallon made little attempt to stay out of the range of Kennedy's punches, hoping for the opportunity to land a brutal left hook or maybe his upper cut. Mallon momentarily dropped his guard ever so slightly. It was the opening that Spike Kennedy was waiting for. He landed a hard left jab on Mallon's jaw. DOWN GOES MALLON! DOWN GOES MALLON!

George Mallon lay motionless on the canvas. The referee

completed the eight-count and Mallon got back up on his feet. The referee checked him and the round proceeded. Moments later, Kennedy put Mallon down a second time. Again, the mandatory eight-count. Then the second round ended. Mallon stumbled back to his corner.

As the third round opened, Kennedy moved in to finish the fight. He unleashed a tremendous onslaught, combination after combination, pummeling Mallon. The Kansas brawler took hit after hit without flinching. He was listless and catching punches, but he showed no panic. Kennedy knocked Mallon to the canvas, once, twice, three times. Each blow made Mallon's head snap back and sent blood spraying. Each time, Mallon rose up after the mandatory eight-count. There were those who did not want to see a fourth round. Mallon was battered and bleeding profusely. The fight seemed to be over. Mallon's corner hoped against hope that he might get in a lucky shot, but did not really imagine that would happen. Some thought the fight should be called. Then there were the fans, some of whom wanted to see more violence and blood. The latter got their wish. George Mallon, sluggish but determined, made his way out from his corner after the bell rang to start the fourth round. The brawler waded into Kennedy, hoping to get in close and land a knock-out punch. The fight was not over until he quit … and he was not going to quit. Kennedy landed yet another blow to Mallon's head. Down went Mallon. He was slow in getting up, but did finally manage to get to his feet. Kennedy continued to mercilessly pound Mallon. There was no parrying. Mallon was no longer slipping any punches. The fight was no longer a contest. There was a long open cut under Mallon's left eye. His right eye was badly swollen shut. His lips were battered and bruised and his nose was split. Blood streamed down his face. Each time Kennedy hit Mallon's face, snapping his head back, blood splattered everywhere. Three more times Kennedy knocked down Mallon. Each time, Mallon refused to quit and with great effort got back up. The bout was no longer a fight, but a slaughter. Not only Mallon, but also Spike Kennedy and referee Dave Porteus were covered with blood. When Mallon bravely rose from the canvas a fifth time, Porteus finally stepped in and called the fight.

The *Junction City Weekly Union* of July 6, 1906 reported:

> The fight was one of the bloodiest affairs that has ever been seen at Kansas City for a long while. Both boys mixed the fighting from the gong and gore was in abundance before the end of the first round. Referee Dave Porteous was nearly covered with blood.

The *Kansas City* (Kansas) *Globe* of July 5, 1906 noted that Mallon:

> ... although apparently whipped from the beginning of the second round, never showed the white feather to the end, which proved bitter to him down to the end of the count in each of the four rounds after the first. To the canvas time after time, Mallon arose again and again and braved his antagonist with the pertinacity and stubbornness of a mad bull, but superior experience and training was too much for the sturdy pipe fitter and he finally submitted to the inevitable after Referee Porteous declared him vanquished and when Armour's killing rooms would present a pale spectacle of gore, compared to his bloody condition.

Despite all this, George Mallon was in the news later that same month. The *Leavenworth Weekly Times* covered a five-fight event at Kirkpatrick's Grove on the Kansas line. An estimated 1,600 spectators were in attendance.

> George Mallon, the heavyweight who trained at the post for a fight with Spike Kennedy July Fourth at Praun's Grove and lost the decision,

was one of the interested spectators! He was very much disappointed when Bates lost his fight, as he and Bates are St. Louisans. Mallon was a policeman in St. Louis when he met Jack Dunleavy and the latter, after boxing with him, took him in charge to make a fighter of him. Mallon knocked out every amateur in St. Louis. He is anxious for a return engagement with Kennedy and has been promised one. Kennedy has a bad hand right now and it will be some weeks before they meet.

The news writer added, "Mallon is one of the nicest and most gentlemanly pugilists in the business and has more friends in the two Kansas Cities than the man who runs a free lunch counter."

The second Mallon-Kennedy fight never took place. It is likely that Kennedy was more focused on working his way up. Kennedy lost to "Great White Hope" Carl Morris in Oklahoma in January 1911 in a fight that received national attention. Mallon missed his opportunity and moved on.

There is no indication that Mallon ever again entered the ring, although he did follow the sport for the rest of his life. On December 22, 1906, just months after the fight with Spike Kennedy, George Mallon married Effie Campbell in Kansas City, Kansas. He was twenty-nine years old. She had just turned eighteen. Perhaps Mallon had concerns about his future prospects in the ring. It seems more likely that Effie was not about to have her husband be a prizefighter. And, if Jack Dunleavy held to his pre-fight statement, he did not encourage Mallon to continue his fight career.

During 1908, George and Effie lived in Kansas City, Missouri where George worked as a fitter for Cunningham Plumbing & Heating Co. The couple lived with Effie's mom and stepdad at 1931 Hallock Drive in Kansas City, Kansas in 1909. Effie's twin sister and her husband also lived in the Hallock house.

George found work in Minneapolis as a fitter and moved north. In September 1909, Effie traveled to Minnesota to join her husband. Thus began a new chapter in George Mallon's life.

George H. Mallon & Effie Campbell Wedding Photo
Courtesy of Diane Gossage

CHAPTER FIVE
MINNEAPOLIS TRADESMAN

By 1912, George and Effie Mallon were living on East Lake Street in Minneapolis. George worked full-time as a steamfitter. Boxing in Minnesota was illegal. It was not until the rise to prominence of St. Paul boxer Tommy Gibbons that the Minnesota Legislature lifted the ban in 1915. By that time, George Mallon was thirty-eight years old and well past a fighter's prime.

In 1913, George went to work for General Fire Extinguisher Company, installing automatic water sprinkler systems in commercial buildings in Minneapolis. The city was a regional center in the early 1900s. The Mississippi River provided the energy to mill the grain being harvested across the Great Plains. As early as 1887, Minneapolis boasted the twelve-story Lumber Exchange Building on 5th Street South. The Lumber Exchange Building was one of the first fireproof buildings in the nation. Frederick Grinnell, an early pioneer in fire safety, formed General Fire Extinguisher Company by merging several smaller companies in 1892. His company was the acknowledged leader in its field. There were numerous outrageous disasters in commercial buildings lacking sprinkler systems in the early 1900s. Among those was the infamous Triangle Shirtwaist Fire, which claimed the lives of 145 in New York City in 1911.

Mallon belonged to Minneapolis Sprinkler Fitters Local 417, which formed ten years earlier. The union hall was at 26 S. Washington Avenue, just off Hennepin Avenue. The union successfully settled a strike and lockout in April 1913.

George Mallon worked hard and was well liked by his colleagues. In May 1913, the sprinkler installers elected Mallon as their delegate to represent them on the Building Trades Council. One article referred to Mallon as "The broad-shouldered delegate from the Sprinkler Fitters...". Life was good for George and Effie. Employers also recognized Mallon's quick mind and his leadership. They did their best to entice him over to their side with compelling offers.

In the summer of 1915, George won the "Fat Man's Race" at the Pipes Trade Picnic. That summer, George Mallon's health issues first began when he was diagnosed with arteriosclerosis. The buildup of fatty plaques and cholesterol leads to hardening of the arteries and risk of a stroke. George Mallon was only thirty-eight years old.

Mallon became friends with Thomas Van Lear, who, like Mallon, was a veteran of the Spanish American War and a business agent for a union. Van Lear nearly won the Minneapolis mayoral election in 1910 and in 1912 as a Socialist candidate. Eugene V. Debs appeared in Minneapolis alongside Van Lear on October 30, 1912 to support his candidacy. Socialism was on the rise nationally. While trade unionism was well established in neighboring St. Paul, Minneapolis remained an open shop town. The key to that was Citizens Alliance, an anti-labor group backed by money from big business in town. Among the key issues at stake was the use of law enforcement to break strikes, rather than to simply keep the peace. The Trades and Labor Assembly, of which Mallon was a member, voted to stand beside the Teamsters during a strike and the corresponding violence with the thugs hired by Citizens Alliance.

In the midst of domestic labor and political strife, events overseas captured the attention of Irish Americans in April 1916. Irish rebels in Dublin rose up on Easter Monday in an armed rebellion, which became known as The Rising. Nothing of such magnitude had taken place in British-occupied Ireland since the days of George Mallon's grandfather and the Rebellion of 1798. Thousands of British troops rushed to Dublin, the center of the fighting. One thousand-plus Irish Republicans put up brave resistance and inflicted heavy casualties on the British. In the end, British artillery prevailed. The Rising lasted only six days. The government brutally oppressed Catholics in the aftermath, executing sixteen and interning large numbers of fighters and civilians in concentration camps.

Brother George H. Mallon, Union Man
Plumbers, Gas and Steam Fitters' Journal
June 1917, page 20.
Courtesy of George Mallon

Later that year, Minneapolis experienced a change in power. Moderates shifted support to the left over the graft and corruption of the right, which was in power. The people of Minneapolis elected Thomas Van Lear as mayor on November 7, 1916. A newspaper praised George Mallon: "He was the hustler of hustlers in the Van Lear campaign as he has always been the hustler of hustlers for everything which pertained to trade unionism."

George Mallon's involvement in Minneapolis politics ceased in the spring of 1917. On April 6, the United States declared war on Germany and prepared to enter the conflict in Europe on the side of the British and French. Mallon was forty years old, a former sergeant and a combat veteran. He answered a call for candidates for officers' training and was accepted into the Reserve Officer Training Corps at Fort Snelling on May 15, 1917.

The *Plumbers, Gas and Steam Fitters Journal* published a letter from M.P. Kavanaugh, Recording Secretary of Local 417, concerning Mallon's enlistment.

> I am forwarding you a picture of Bro. Geo. H. Mallon, business agent of Local 417, Sprinkler Fitters and Helpers, located in Minneapolis. Local 417 requests that you publish the picture and these few lines in the next copy of the Journal to let friends and members, in general, know of the enlistment of Bro. Mallon in the U.S. Army in response to the call of his country. He is now in the reserve officers training camp stationed at Fort Snelling. Local 417 feels the loss of Bro. Mallon very much, as he has always been a loyal brother and a very competent business agent, handling matters of great importance in a very satisfactory and business like way for our Local and those concerned. Bro. Mallon was also a member of the Minneapolis Building Trades Council and the Twin City Pipe Trades Council. He always had the welfare of the U. A. and organized labor first in his thoughts. Bro. Mallon is a veteran of a former campaign, having served three years in the regular army and we know that where he goes in the future he will be found as he always has in the past – with his face to the foe.

While we feel the loss of Bro. Mallon very much, we are also proud of him and when this war is over we sincerely hope and trust to have him with us again to take up his duties where he left off to serve his country and the U. A. and Bro. Mallon may rest assured he will receive a royal welcome.

Mallon, for his part, assured his friend Mayor Van Lear and all his colleagues in the labor movement, "I will be just as good a union man now as ever and I hope I will be back to work. For Labor and Van Lear again at the next election."

CHAPTER SIX
GOING OVER THERE

The United States maintained a policy of neutrality through the first three years of the war in Europe. Woodrow Wilson was re-elected president on November 7, 1916 with the campaign slogan "He kept us out of war." A subsequent series of surprising moves by German poisoned its diplomatic relations with the United States. On April 2, 1917, President Wilson told Congress:

> The world must be made safe for democracy. Its peace must be planted upon the tested foundations of political liberty. We have no selfish ends to serve ... We are but one of the champions of the rights of mankind. We shall be satisfied when those rights have been made as secure as the faith and the freedom of nations can make them.

On April 6, 1917, the United States Congress declared war on Germany.

The United States Army was unprepared for war. The situation was far different from going against Spain in 1898. Germany was a colossal military power with leading-edge weapons. Concern over the Zimmerman Telegram was the sense that Mexican cooperation with Germany could result in veteran German divisions invading through the southern border of the United States. The United States began opening camps and commencing mobilization and training. Citizens responded with a surge of patriotism, volunteering to go to war.

On August 15, 1917, the United States Army commissioned George H. Mallon as Captain of Infantry. Captain Mallon received orders to proceed to Camp Logan in Houston, Texas. The new military installation under construction was to become the training camp for the Illinois National Guard. George Mallon's wife Effie moved back to Kansas to be with family. Mallon arrived at Camp Logan on September 6. The situation at Camp Logan and in nearby Houston was tense when Mallon arrived.

Prior to Mallon's arrival, the Army moved the 3rd Battalion of the African American 24th U.S. Infantry to guard the construction going on at Camp Logan. The regular army veterans experienced no Jim Crow treatment at its previous posting in Columbus, New Mexico. There was considerable concern that the Houston police might provoke an incident, given their attitude toward the newly arrived African American soldiers, who were not sufficiently submissive.

Racial tensions already were on edge throughout the nation because of rioting in East St. Louis in July. White mobs murdered an estimated 100 to 200 African Americans and torched entire neighborhoods, leaving thousands homeless. At issue was the migration northward of Southern African Americans and their employment in the wartime industries. Ten thousand African Americans marched silently in protest down Fifth Avenue in New York City. African Americans across the nation contributed to the re-building of East St. Louis. Some of the soldiers of the 24th U.S. Infantry were among the contributors.

There were minor incidents in Houston, beginning soon after the arrival of the African American soldiers. Camp Logan was situated just two miles west of Houston. On the westernmost edge of town was the Fourth Ward, site of what was known as Freedmen's Town. The settlement was located on the southern bank of Buffalo Bayou, which was prone to flooding and undesirable to whites. The soldiers on leave mingled freely with local African Americans. There was concern among the authorities about the soldiers' posture causing unrest among local African Americans.

On August 23, just days before Mallon's arrival at Camp Logan, the police triggered an incident. Two officers broke up a crap game on a street corner. In pursuit of the suspects, they forced their way into the house of an African American woman, struck her and dragged her outside. Private Alonzo Edwards of the 24th approached the officers, who pistol-whipped, then arrested him. When Corporal Baltimore of the 24th later inquired as to the status of Private Edwards, the officers struck the corporal and then opened fire on him as he fled. They caught up with the corporal, beat him badly and placed him in jail with Edwards. In order to calm unrest back in the camp, a white officer of the 24th managed to retrieve Corporal Baltimore. While that did show that the corporal was still alive, the corporal's badly beaten condition roused the soldiers. A rumor that a mob was marching on the camp led to most of one company and more disregarding their officers, mobbing the supply tents and arming themselves. Sergeant Vida Henry led some 150 angry soldiers in a two-mile march into town. An intense firefight commenced as they reached town. They shot and killed three among a group of police officers, one of whom happened to be one of the arresting officers from earlier in the day. In the midst of the chaos, a car approached the soldiers. The soldiers opened fire, only to discover that they killed Captain Joseph Mattes of the Illinois National Guard. The seriousness of this caused many of the mutineers to begin to drift off. In all, five police officers, nine civilians and four soldiers died in the fight. One of the soldiers killed was Captain Joseph A. Mattes of the 33rd Division.

By the time that Mallon arrived at Camp Logan, the 3rd Battalion of the 24th Regiment was back in New Mexico. A military tribunal indicted 118 enlisted men of Company I. The first of three court-martials began at Fort Sam Houston in San Antonio in December. The authorities sentenced and quietly hung 13 soldiers, then buried them in a non-disclosed location. As with most riots, it is likely that the tribunal was not certain of the duplicity of all, whom it sentenced to death. In September, the U.S. Army hung 6 more soldiers. 63 others received life sentences in prison. The civil court in Houston brought no white police officers or civilians to trial. Despite such treatment, more than 350,00 African Americans served in the U.S. Army during the

World War. In contrast, many Catholics in the northern region of Ireland resisted serving in the hated British Army in what Europeans referred to as the Great War.

Captain Mallon was assigned to the headquarters of Brigadier General D. Jack Foster of the 66th Infantry Brigade, 33rd Division, for special duty. The United States Army activated the Illinois National Guard in July 1917. They were known as The Prairie Division. Elements of the 33rd Division were from Chicago, which resulted in considerable diversity in the unit. The 132nd Regiment's chaplain was Captain John L. O'Donnell, assistant pastor of St. Patrick's Church in the West Loop of Chicago. Both the Mallons and the O'Donnells traced their lineage back to County Tyrone. The Houston Post of October 22, 1917, reported that Father O'Donnell said Mass every Sunday at 8:00 a.m. and over eight hundred attended. The article added that "… more would have been there had it been possible to crowd inside."

A situation developed on the Texas Coast which was contrary to all that George Mallon stood for. Poor working conditions and low pay led to oilfield workers to form a labor union at Goose Creek, east of Houston, in late 1916. They affiliated with Samuel Gompers' American Federation of Labor (AFL). Oil producers refused to even discuss grievances. As a result, 10,000 oilfield workers in Texas and Louisiana went on strike on November 1, 1917. Oilfield workers also went on strike in the California oil fields. Producers claimed that the organizers of the radical International Workers of the World (IWW) instigated the strike. What really made the situation seem ominous was that oilfield workers throughout Mexico were going on strike. Memories of the Zimmerman Telegram, which proved German interest in Mexico, lingered in the minds of Americans. Rumors spread that Germans were trying to interfere with fuel for the Allies.

The governors of Texas and Louisiana, who were aligned with the big money interests, appealed to the Secretary of War Newton Baker for troops to protect the oilfields. Within two days, the commanding general of the Southern Department of the U.S. Army began to deploy troops. There were some tentative moments, but no violence. The Army reported that, if necessary, the oil companies would bring in nonunion labor under the protection of the troops. The entire situation was troubling and painful to any soldier, who was a union man. George Mallon knew that he had no option but to obey orders as any soldier must.

The matter became even worse when First Battalion, 132nd Regiment, was assigned to strike duty at Sour Lake Oil Field, seventy-five miles northeast of Houston. The discovery was a major find for Joe Cullinan and The Texas Company, predecessor of Texaco, in 1913. On the night of November 7, a weapon accidentally discharged in a tent. The round went off into the night. No one heard a cry or a groan. An hour later, someone passing between tents came across the body of Captain Oscar Hogstedt, commanding Company D, 132nd Regiment. The same company lost a commanding officer in an accident in the Philippines. The 33rd Division commander sent a lieutenant colonel and Chaplain John L. O'Donnell to investigate and make arrangements to ship the body home to Chicago.

The Prairie Division underwent rigorous training in early January in preparation for shipping overseas to enter into the fight. Field hospitals and ambulance sections participated in two-day war maneuvers. Stretcher bearers carried "wounded" from first aid stations near "the front" to dressing stations, where ambulances drawn by mule teams took them to field hospitals. Units moved forward in the dark of night behind the advancing front, taking all precautions. They also made a rapid evacuation when word came that the enemy was advancing. Men hiked twelve miles in the course of four hours. Everyone underwent training with automatic rifles, both the French Chauchat and the American Lewis. Army inspectors went over business places near the camp. Soldiers were not allowed to visit any establishment that did not have a certificate of license posted.

The quarters at Camp Logan consisted of canvas tents.

On January 10 through 12, 1918, the temperature fell to 11 degrees and Houston was blanketed with heavy snow. George Mallon of Minneapolis experienced the Great January 1918 Blizzard.

The *Houston Daily Post* of January 15, 1918 reported, "Forty Officers at Camp Logan Are Promoted. From the Officers' Reserve Corps: To be captain, G.H. Mallon ...". On the basis of his previous military experience, he was given the command of Company E, 132nd Infantry Regiment, 33rd Division. The core of the Prairie Division, as the 33rd was known, was comprised of men from the Illinois National Guard.

Captain Mallon's first sergeant was Sidney Gumpertz. Like Mallon, he was older. Gumpertz was thirty-nine years old, above the draft age range. He also was married. When a reporter asked Gumpertz why he enlisted, the sergeant replied, "... because I am a patriot. I don't know why but I always have been." More than one newspaper account referred to the pair as "the Irish captain and the Jewish sergeant." Gumpertz described Mallon, as "a big, iron-fisted, square-jawed Irishman from Minnesota, with steel-blue eyes and a steel-blue will." Mallon, for his part, considered Gumpertz, "the most loyal and thoroughly efficient fighting man I have ever known." The two got along well and forged the company of recruits into fighting shape. Gumpertz proudly recalled, "We had tough bayonets in our company." General Pershing, commanding the American Expeditionary Force, insisted that training emphasize marksmanship. Sergeant Gumpertz said, "I was a pretty good shot. Not as good as Mallon. He always got the bull's eye ... I didn't make many bull's eyes, like Mallon, but I was always shooting around the bull's eye; well, within the breadth of a man."

A newspaper reporter from Kansas City wrote a small feature on Captain Mallon of Kansas City. The title was "Captain Mallon Working Hard to Improve Co. E." The article stated that the discipline and the efficiency of the company had increased one-hundred-percent since Captain Mallon had taken over. The article credited Mallon's leadership for motivating his men through his personality. The description of the officer glowed:

> Captain Mallon is a very fine specimen of manhood; tall, well-built, at one time light and heavyweight champion, he appears as a great Greek warrior ... he says his life began when he became commander of Company E, 132nd Infantry.

Mallon liked and understood people. He was not one to use coercion to implement discipline among his soldiers. He earned their respect and solidified his relationship with the men as the one whom they would rely upon to lead them to victory and get them home.

The 33rd Division left Camp Logan on May 5, 1918. They traveled by train to the port of Hoboken, New Jersey. On May 16, Captain Mallon and his two-hundred-and-fifty men boarded the Mt. Vernon and shipped off for France. Captain Mallon and his men arrived at the French port of Brest on May 26 and became part of General John J. Pershing's American Expeditionary Force. They would write history.

CHAPTER SEVEN
BATTLE OF HAMEL

In June 1918, the 132nd Regiment was stationed at Allery, France for additional training. They no longer marched to the cheers of thousands, as they did when they departed. Instead, they slogged through mud and learned about trench warfare. However, there is no better training than actual warfare. Most of Captain Mallon's 250 men had never experienced combat. It had been more than fifteen years since Mallon saw combat.

The Germans were extremely concerned with the monumental shift in the war as American soldiers crossed the Atlantic Ocean to France. It was to the German advantage that Lenin and the Bolsheviks took Russia out of the Great War. The Germans shifted many divisions to the west. They launched a series of major spring offensives in 1918 in the hopes of seizing victory before the Americans shifted the balance in the conflict. The first victory for U.S. troops took place at Cantigny on May 28, 1918. American soldiers stopped the German advance on Paris at Chateau-Thierry in late May and early June. The Marines took Belleau Wood from the Germans as the end of June approached. Most of the American Expeditionary Force, including Captain George Mallon and his Company E of the 132nd Regiment of the 33rd Division, had yet to see action.

Among the many German gains from their spring offensives was a salient on the south bank of the Somme River near the major French city of Amiens. The high ground, which the Germans held, allowed their artillery observers a commanding view from which to target British positions. The Germans held that salient since April 4, 1918. The Allied high command was very concerned about the situation, which they considered a grave danger.

Captain Mallon Doughboy Hero

Captain George H. Mallon, 132nd Infantry Regiment
while deployed in France
courtesy of Diane Gossage

In June, Australian Lt. General John Monash conceived of a plan to take Hamel and eliminate the salient. The success of the operation also would establish a jumping off point for future Allied offensives. Monash's challenge was that the ranks of his Australian Corps were much depleted from the continual combat of months in the line.

General Monash presented a request to General Sir Henry Rawlinson, commanding British Fourth Army, to incorporate untested American units into his Australian units. That would not only augment his ranks, but also allow the Americans the opportunity for some operational experience. Rawlinson approved. He sent a request to the Major General Read, the commander of the A.E.F. II Corps, who passed on an order to Major General George Bell, Jr., commanding 33rd Division, on June 27, 1918. General Bell chose two companies each from Companies A and E, 132nd Infantry Regiment. When the announcement was made, there was great cheering by the men being sent out.

On the morning of June 30, four American companies were inserted into four Australian battalions. Company C, 131st Infantry Regiment, was assigned to 42nd Battalion from Queensland. Company E, 131st Infantry Regiment was assigned to the 43rd battalion from South Australia. Company A, 132nd Infantry Regiment was assigned to 13th Battalion from New South Wales. Company G, 132nd Infantry Regiment was assigned to the Queensland 15th Battalion. Then each American platoon was assigned to an Australian company. That allowed the green American doughboys to serve alongside veteran diggers, as the Aussies were called.

There was some concern about the Americans suggesting they had come to win the war. A digger reportedly asked of them, "Are you going to win the war for us?" A doughboy answered, "We hope we'll fight like the Australians." A strong camaraderie soon developed. The fresh spirit and enthusiastic attitude of the Americans doughboys buoyed the diggers, who had fought so hard for so long and suffered the loss of so many comrades. One

Australian officer reported of the doughboys, "From the first when our soldiers came in contact with them, they mixed well and took kindly to each other." Another wrote, "These American soldiers seem an exceptionally smart lot and are most keen to learn. They are getting on exceptionally well with the officers and men of the brigade." A war correspondent wrote, "One of their officers told Wilkins the other day that he felt as much at home amongst Australians as amongst his own countrymen."

Regimental command assigned Captain Mallon and Sergeant Gumpertz from Company E, 132nd Company to accompany the soldiers of Company G, 132nd Infantry with the Queensland 15th Battalion. Regimental Chaplain John O'Donnell also accompanied the men of the 132nd Regiment. The story as to how exactly Mallon and his buddy Gumpertz managed to be inserted into the contingent going to the Aussies is not known, but, no doubt, interesting. All in all, twenty-five American officers and 1,244 enlisted men joined the Australians.

The attack was to commence early on the morning of July 4, a fitting day for the doughboys to commence their battle experience in Europe. A major parade was scheduled for that same day in Paris with General Pershing and select AEF soldiers involved.

The American doughboys slipped into the forward positions on the nights of July 2 and July 3. General John Monash, unlike too many of his peers, recognized the insanity of needlessly expending the lives of his soldiers in unsupported frontal assaults against entrenched positions laced with machine guns. He devised a combined arms operation. By summer 1918, the British had new tanks in the field. The Mark V tanks were not only better armed and more potent weapons, they were also much faster and more maneuverable. The tanks were to attack in three lines. The first line was to precede the infantry and act independently. It was their mission to bust open pathways through the barbed wire, destroy key defensive positions, assist in protecting the infantry and breakthrough into the rear of the enemy position. The second line of tanks alongside the infantry was to protect the infantry and to support the infantry in mopping up. The third line was composed of tanks to replace those lost in the first and second lines, as well as to eliminate any and all final resistance after the main wave passed through. Monash placed the tanks under control of the infantry officers for the specific purpose of minimizing infantry casualties. The tanks were to closely follow a rolling artillery barrage and eliminate bunkers and machine gun nests. The infantry would follow closely, using the tanks as cover, and mop-up what remained. In all, sixty tanks were committed. The assault was to commence in the early morning of July 4.

Early on the morning of July 3, Pershing discovered that some of his doughboys were going to be seeing action under Australian command. He was not at all happy and ordered the Americans to withdraw immediately. Company C, 131st Regiment did so. Others, including Company G, 132nd Regiment with Captain Mallon did not respond to the order. The doughboys were anxious to get into the action.

At 4 p.m. on July 3, General Rawlinson, under pressure from his superior, British Field Marshal Douglas Haig, ordered General Monash to withdraw all Americans from his line. Monash, at considerable risk, insisted to the contrary. He argued that, while he was prepared to comply with the direct order, that the plans were already in place and the units under question were already in position. Rawlinson, at considerable risk, went back to Haig, telling him that withdrawing the Americans would cause abandonment of the entire operation. Haig conceded. He left Pershing out of the loop. It was too late to stop them.

The Americans settled into the trenches to await the attack. There was some anxiety about gas. "As none of us had ever smelt gas, we were not sure what it smelled like." And a myriad of smells wafted over the trenches from the surrounding battlefield.

Under cover of darkness, infantry squads went out and cut the barbed wire and engineers laid tapes across the field of advance through the openings. The assault force quietly made their way into the waist-high wheat fields, each taking resting positions behind the beginning of its respective tape. While not standard for the Australian Corps, each soldier received a tot of rum to warm his insides and calm him. An occasional German flare lit up the sky and fell to the ground in front of them. Allied planes passed over and harassed the German lines. The air operations not only kept the Germans from sleeping, but also masked the sound of tanks slowly idling into position. The Americans with the 15th Battalion – Captain Mallon and Company G, 132nd Infantry – were distributed in sections, rather than in platoons, contrary to orders. That spread them out more among veteran soldiers.

The Allied artillery commenced at 3:00 a.m. on July 4. It was a Fourth of July fireworks display like Captain Mallon had never before seen. At 3:10 a.m., the artillery shifted to a rolling barrage. Every second shell was shrapnel, mixed with smoke and high explosives. Precision artillery did not yet exist. Some of the guns were old with worn barrels. In several places along the line, some shells fell short, resulting in tragic loss to friendly fire. Company E, 131st Infantry lost a squad. Captain Mallon and the men of Company G, 132nd Regiment were just moving out when some artillery shells fell 300 yards short. Chaos reigned. Sounds became muted. Everything occurred in slow motion. The images of the dead and dying and wounded were intense and, likely, forever embedded in memory. When the smoke settled, twelve men of Company G, 132nd Infantry lay dead with another thirty wounded and screaming in agony.

No doubt in shock, Captain Mallon and the able-bodied of Company G moved forward with Queensland 15th Battalion. The risk of loss to friendly fire remained, given the overeager doughboys determined to stay up with the diggers in the advance. At least one Australian officer lost his life running ahead to pull back American soldiers getting too far ahead. Mallon later wrote of this challenge for officers. Meanwhile, the smoke, combined with the darkness, caused considerable confusion as to orientation and direction.

They encountered strong resistance at their initial objective, the Pear Trench. A substantial patch of concertina and barbed wire remained intact. As the assault teams stacked up, machine guns raked their fields of fire. It was a tense moment. No one knew if they would survive. Tanks were still moving up. Aussies opened up with their Lewis guns, firing from the hip, doing their best to suppress and perhaps eliminate the machine gunners taking their toll of the attackers. Meanwhile, other units flanked the Pear Trench and the center rose up and made its way into center of the German position.

Pear Trench was crowded with Germans and machine guns. The position was seventy yards deep. The diggers and doughboys poured in. Some Germans surrendered while others kept fighting. It became a killing field. Captain Mallon observed forty dead Germans in a very small sector. The diggers and doughboys moved on to catch up with the rolling barrage. There was no formation. Men just moved forward in small groups. They reached their halt-line and re-formed under a heavy smoke screen and waited for the tanks to catch up. The surge resumed at 4:10 a.m. The Germans were in well entrenched positions beyond the Pear Trench and made a determined stand. Fighting was close and fierce. It was only through the fearless gallantry of the spirited Aussies and Americans that they took their final objectives.

Mallon's after-action description was concise:

> ... first objective ... took about fifteen minutes to reach ... a great many Germans killed ... We also lost heavily in front of two machine guns before they were silenced by troops attacking from the rear after surrounding the machine guns ... the first wave passed onto a ridge, about a thousand yards ... captured a great number of prisoners ... the tanks, three on each flank, did very good work.

The doughboys, their adrenalin soaring, cheered upon finally driving off the Germans and taking their final objective. For many of them, it was their baptism of fire. Reality returned when the diggers, veterans of too many bloody fights to recall, moved quickly to dig in and prepare for the inevitable counterattack. It was 7:00 a.m. The assault on Hamel achieved its objectives in ninety-three minutes.

The Germans shelled the position throughout the day. Tanks brought up supplies, a far more efficient and safer alternative to exposing men crossing the ground on foot. The Allies controlled the air. Lt. Gen. Monash's plan combined arms in every way. Aeroplanes dropped ammunition cases by parachute to the newly established front line by parachute.

Around noontime, thirty-five German warplanes appeared. Captain Mallon and his comrades watched in astonishment as a flurry of dogfights took place in the skies above them. Two Allied planes went down in the melee.

At dusk, the Germans hit the newly-taken positions with a fierce charge. Captain Mallon and Company G took part in a bold counterattack that shocked the Germans and broke their charge. The Allies took fifty prisoners in the action.

> The men immediately after reaching the objective front line dug in immediately and prepared for the counterattack, which did not take place until the evening ... All day of the 4th they shelled out trenches ... Our battalion was relieved in the night of the 5th and reported back to the regiment on July 6th.

Captain Mallon's report made no mention of his own actions. Mallon's First Sergeant Sydney Gumpertz provided that.

They were nearing a trench; Gumpertz's automatic was spitting, but Mallon was holding his idle in his hand. They came to a German officer and Mallon, still disdaining his automatic, brought up a sharp left against the enemy jaw and knocked the German out … Mallon leaped into the trench and felled the two Germans with the same fist that had done for their officer.

Captain Mallon, it seemed, retained a preference for and confidence in his once-famous right uppercut. Of course, the well thrown punches subdued the German soldiers without killing them.

American General John J. Pershing was more than a little surprised when he learned that his doughboys took part in the fight at Hamel. He praised the American performance. He also made very clear to all under his command that nothing of the kind would take place again. It did not. The Battle of Hamel was important, beyond reducing the salient, in demonstrating again the fighting spirit and capability of the American soldier.

Lieutenant General Sir John Monash, legendary commander of the Australian Corps, lauded the American soldiers for their "dash, gallantry and efficiency." That surely applied to Captain George Mallon. Many stories came out of the British and Australians, who had their first combat experience with the doughboys. One remark heard was, "You'll do me, Yank, but you chaps are a bit rough." An Australian colonel reportedly said, "Yanks, you are fighting fools, but I'm for you!"

The British decorated fourteen Americans. The American Army awarded eight Americans soldiers with the Distinguished Service Cross and one American with the Medal of Honor for their valor at Hamel. Mallon was not among them.

CHAPTER EIGHT
MEUSE-ARGONNE OFFENSIVE

The Meuse-Argonne Offensive was the final event of the First World War. It remains the largest battle in American military history, involving 1.2 million American soldiers. The objective was to sweep clear of Germans the entire area between the Meuse River on the east and the Argonne Forest on the west. The ultimate strategic objective, thirty-five miles to the north of the American lines, was the rail junction at Sedan. All lateral communications and supplies between German forces east and west of the Meuse River ran through Sedan. The challenge was that the terrain was ideal for defense and the Germans occupied nearly continuous defensive positions in depth for more than ten miles. The Germans had four years to dig in and emplace guns. The high ridges overlapped, providing the Germans with excellent positions for observation and converging fields of fire for killing zones.

Company E, 132nd Regiment made a series of forced night marches to reach Le Mort Homme (Dead Man's Hill), their assigned position at the front. Le Mort Homme was northwest of Verdun and had been the scene of many deadly battles. Captain Mallon and his men slogged through muck and mud down nearly impassable roads jammed with trucks, which had sunk to their axles. Column after column of tramping men made their way down the road. Mallon, frustrated by the near gridlock, finally led his men off the roads and cross-country through the scarred landscape. It was difficult work, cutting barbed wire in the dark. The 132nd men moved into the secondary line just off the crest of Le Mort Homme at the end of the first week of September. Shortly thereafter, the 132nd moved down the slope, relieving the French troops in the jumping off line. During this period, the doughboys actively patrolled No Man's Land and saw considerable

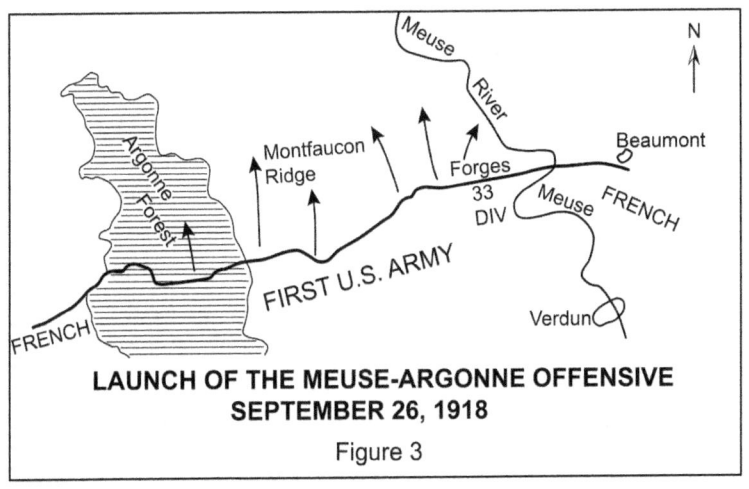

**LAUNCH OF THE MEUSE-ARGONNE OFFENSIVE
SEPTEMBER 26, 1918**
Figure 3

action in small unit encounters with German night patrols.

Everyone on both sides knew that the buildup was a prelude to the American's big push. By September 24, the latrine rumors were widespread that the offensive was about to commence. Orders came through on September 25. At 2300 hours on September 25, Captain Mallon and his regiment moved through communication trenches to the front line. Thick, dense fog muted sounds and, mixed with the black of night, lent an eerie air to the silent passage of the long column of soldiers. Their boots splashed and sank in the muck and mud, leaving everyone feeling miserable in their wet socks. Mallon's Company E was in position by 0325 hours in the early morning of September 26, 1918. They were on the battalion's left in the Massin Trench. The German positions lay across the valley. The captain sent out small parties to cut the wire in front of their trench. As they returned to the line, their ghostly figures emerged out of the fog when they were only fifteen or twenty feet away.

The ground before the 132nd Regiment/33rd Division was challenging. The mission was to cross Forges Brook, break through enemy positions in front of and in Bois de Forges and organize on the Verdun-Sedan Road, 400 meters north of Bois de Forges. The first obstacle was Forges Brook, which separated the Americans from the German defensive line. The brook was narrow, but the valley was open and marshy. Some reports referred to the valley as Forges Swamp.

A deafening artillery barrage commenced at 0230 and continued for three hours. The 33rd Division's 108th Engineers quietly slipped over the top and disappeared into the fog. It was the combat engineers' mission to prepare the way through the swamp and across the brook to the German positions on the far side. They had spent hours and hours in preparation for that moment. The engineers carried forward 12,000 fascines – bundles of sticks bound together – and wooden planks and other building materials. They cut paths through the tangle of barbed wire and then moved toward the brook, which ran through the center of the swamp. The Germans, unable to identify specific targets, raked their fields of fire with machine guns and lobbed artillery shells into No Man's Land. The engineers calmly constructed nine passageways across Forges Brook. The engineers then taped off V-shaped avenues to lead the assault forces to each of the bridges.

Captain Mallon's focus was on his men. He had faith in his first sergeant, Sydney Gumpertz. The two of them had trained the men well. It was up to them to lead the soldiers of Company E. If they led well, the men would follow. Mallon showed his men in training that he had genuine concern for their well-being and that he was willing to share hardship with them. He also showed them that he accepted full responsibility for any consequences, but attributed success and accolades to the men under his command. This was the art of leadership, which George Mallon learned as a non-commissioned officer in the dark jungles of the Philippines.

The Germans awaiting the 33rd Division were veterans of countless battles and well positioned on good ground. There would be casualties among the men under Captain Mallon's command. The only question was how many. Mallon felt that burden of responsibility for the lives of more than one hundred and fifty good men.

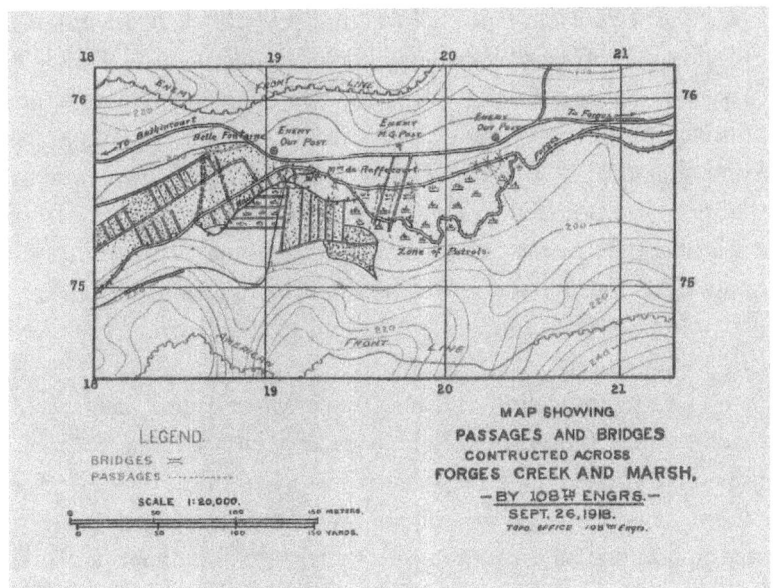

Map of the objective, Forges Creek
132nd Regiment, 33rd Division
September 26, 1918
from Huidekoper's History of the 33rd Division

Mallon surveyed the fog that enveloped the field of battle. He acknowledged that the limited visibility of the fog would make command & control difficult. That would only add to the chaos of battle, which many of his men were about to experience for the first time. The captain also recognized opportunity. The fog would screen the movements of the assault teams from the German machine guns and artillery observers. That seemed providential.

The order was to advance at 0530 hours. Captain Mallon and his Company E were in the first wave. As the pending moment approached, Captain Mallon climbed out of the trench to take his position at the head of his company. His adrenalin was pumping. He made every effort to look calm and composed for his men. At that moment, the Germans let loose an artillery barrage. Mallon stood firm and resolute. His men were poised, ready to go over the top. He wondered to himself if anyone would be able to hear the command to advance.

At 0530, Captain Mallon raised his whistle to his mouth. He and other officers blew their whistles simultaneously. The shrill blasts cut through the fog. Company E, along with so many other companies, went over the top and surged forward into No Man's Land. Regimental chaplain Father John O'Donnell was among the men of Company E. His eyes were nearly swollen shut from a gassing. The priest carried a Colt 45 automatic pistol in each hand. Sergeant Sydney Gumpertz and his platoon disappeared off to Mallon's left in the fog. A rolling artillery barrage preceded the assault troops.

Captain Mallon and his men advanced into the face of machine gun fire and exploding artillery shells. They followed the tape to the edge of the swamp and a makeshift plank bridge. Mallon saw fingers on the edge of the plank. The engineers were standing in the water, some up to their necks, holding the planks in place. Captain Mallon and his men trotted across the bridge, each step tentative. The infantrymen streamed onto the far bank while under heavy artillery and machine gun fire. The rat-tat-tat of machine gun fire intensified from a mass of tiered machine gun nests before them. The woods were thick and the dense fog hung low. Mallon later related, "The fog made the battle one between blind men." Mallon was comfortable with tactically maneuvering through the haze to achieve his mission.

Men became separated and lost in the chaos. Individuals emerged from the fog and disappeared as quickly. The machine gun fire added to the chaos of the moment. Small groups moved forward. Captain Mallon and twenty doughboys scrambled toward the flash of a Maxim machine gun. He had a Colt .45 Automatic pistol in one hand and a grenade in the other. They advanced in a broad skirmish line. The fog hid them, but they were very vulnerable to the random sweeps of the machine gun bursts. As they neared, a soldier to Mallon's right lobbed a hand grenade at the flash. The explosion ripped through the fog and the gun went silent.

Mallon knew to mentally distance himself from the chaos so he could make good decisions. He led his band forward into the German complex of machine gun nests. They used the gun flashes as their guide. Barbed wire was everywhere. Each of them had wire cutters and did his part to cut open a path while under fire. Machine gun bursts send bits of dirt, rock and wood into the air. Mallon led his men to flank the nest. He ran up the parapet and leaped down into the nest. He dropped the startled gunner with a left to the chin. One of his men dispatched the other gunner. The other Germans raised their hands in surrender and cried, "Kamerad!" Other Yanks came up and took the prisoners. Captain Mallon and his small band moved on.

The woods were thick and full of machine gun nests. Mallon and his men continued the tactic of rapidly appearing out of the fog and rushing each one. Some Germans resisted and died. Some managed to flee. Others surrendered. Mallon later remarked that they, "... found Hun hunting good."

In one clearing, Mallon was running forward when his knees buckled and he fell hard to the ground. The arms of a German soldier were wrapped around his legs. Mallon raised his pistol to crash it into the Hun's head. "I stayed the blow to get a better look at my assailant." It was a boy. Captain Mallon later recalled that the boy's face, "... reflected a terror and pleading for life that I hope I never see again in a human." He patted the boy on the head, picked him up and sent him to the rear on his own recognizance. "I saw many more youngsters that day trying to do a soldier's work of maiming and killing. Fortunately, I did not have to use my gun on them."

A succession of brilliant red muzzle flashes and earth-shaking booms caught Captain Mallon's attention. Big guns. Probably howitzers, sending death and destruction into the American ranks. Mallon later explained his decision process as a field commander:

> Like a flash came the words, which my instructor, John F. Franklin of the Regular Army, used to say to our class at Fort Snelling when we were in training. He said, "When in battle and in doubt

as to what to do, do something, push ahead." So, the first thought that came into my mind was the command to charge. I yelled it to the men and then the big job for me was to keep up with those Illinois boys. It was the old Illinois National Guard, you know. Well, using our rifles and bayonets and automatics, we captured those four big guns. Don't ask me about details. It all happened so fast and we were all at it so hard that it's pretty hard to explain exactly what happened.

Captain Mallon had eight men left. They had no idea how many Germans they would be facing as they scrambled toward the red flashes.

A similar version in Mallon's words:

... All I could think to do was to charge and we went at them with bayonets and pistols and we were too close for them to shoot at us. It took only three or four minutes to clean them up for most of them made a "strategic retreat". A few held up their hands and the rest of them are there yet. I expect if I had more time to think I would have started the other way as they numbered ten to one and if they had our nerve there would be a different story to tell.

Another version of the story, which Mallon told, contained additional details:

... it was a surprised bunch of Germans who saw nine, battle-torn Yanks come sailing over the earthworks in front of the emplacement ... A Prussian officer about my size, and I'm over six feet tall, met me halfway to the bottom with half-drawn pistol. There was no chance to argue and

I let him have one full on the chin. There was an automatic in one of my hands or the other, but I'm not sure which I hit him with, fist or gun. Whichever it was, it had the desired effect and the Hum went sailing end-over-end back into the pit and nine Yanks swarmed after him. I always will contend that a good solid fist inspired more fear and obedience in the Germans than did anything else unless it was naked steel.

What Mallon did not specifically state was that he was able to capture the battery without having to kill the German officer. That clearly was important to him. For his bravery in the above-described incident, Captain George Mallon received the Medal of Honor.

Captain Mallon caught his breath and glanced about at his boys. They had done well in their baptism of fire. Corporal Harry Dawson and Private Vivian Badger were solid and deserving of promotion. Two others were not even citizens. Anton Churas was Lithuanian and Sam Salpietro was Sicilian. Rugged warriors. Then there were Privates Allen Bridges and Albert Wittman. All of them were tough kids from Chicago. There was also Private William Kelly, a farm kid from Nebraska, and Private Gustave Schill from Pennsylvania. American boys now warriors.

Captain Mallon and his small band of doughboys captured four massive 155mm howitzers, an antiaircraft gun and a number of crated and uncrated Maxim machine guns in addition to "enough prisoners to fill a good-sized prison pen." While trying to decide how to handle his large catch of Germans, Captain Mallon was relieved to see First Sergeant Gumpertz suddenly appear with the remnants of a platoon.

Mallon's First Sergeant, Syd Gumpertz, led his platoon of fifty men into the fog and in the face of artillery fire and machine gun fire. At some point, Gumpertz slid into a trench. His boys began to follow him. Gumpertz told them to get out of the trench and then tossed a smoke grenade. Smoke filled the crevices and dugouts. In short time, some fifty Germans, gasping for air, poured out and surrendered. Gumpertz sent some men back with the prisoners and continued forward. They came across a machine gun, hidden in front of them, spraying the area with a hailstorm of lead.

Gumpertz asked for two volunteers to assist him in taking out the machine gun nest. Corporal Paul Siclair, a French-Canadian, and Private Sebastian Emma, an Italian, stepped forward. Gumpertz considered Siclair a model soldier, but he had his reservations about Emma, who was careless with his equipment and too often AWOL. No time to re-consider, Gumpertz motioned for them to follow and they headed out. They did not crawl or even try to flank the machine gun nest. They simply raced into the fog toward the sound of the gun. They ran frantically, not sure if they might meet their end any second. They suddenly felt the ground rise up and realized, as they continued up, that it was the slope of a parapet, part of the machine gun emplacement. Gumpertz nearly fell into the position. They looked down to see the crew firing away. Gumpertz opened up with his automatic pistol and the other two with their rifles. Gumpertz unloaded his chamber on the Germans. At the end of a brief firefight and a few dead Germans, the others surrendered. Their prize was a heavy machine gun and fourteen prisoners. They moved on as the rest of the platoon came up. Minutes later, a shell landed in the midst of the three men in the vanguard of the platoon. The sergeant picked himself up off the ground to find Corporal Siclair dead, killed outright. Gumpertz crawled over to Private Emma, who was badly wounded. Gumpertz knew he had been hard on the little Italian. He told him, "That's all right, Emma. You're going to be all right." Emma looked up at him and replied, "Oh, no, Sergeant. I die. I die. Goodbye." Then the young hero's head slumped and he passed away.

Sergeant Gumpertz changed out the clip on his automatic pistol and went on alone, not waiting for the platoon to come up. More machine gun fire was raining in on his platoon through the fog. Bullets zipped through his pants and his shirt, just missing him.

He ran up another parapet, still in the fog, and suddenly emerged out of the fog over an active machine gun and its crew. Gumpertz shot the gunner and the man next to the gunner with his pistol. He tossed down a hand grenade that exploded with deadly results in the trench. He shot another man. The rest of the men raised their hands. Gumpertz accepted the surrender of sixteen prisoners. In the nest were found two heavy machine guns.

Captain Mallon, reunited with Sergeant Gumpertz, left a few of the doughboys to guard the prisoners and the howitzers and moved forward, having combined their bands. There was more fighting that day and in the days to come. Like his friend Captain Mallon, Sergeant Gumpertz continued to lead the advance and place himself in harm's way.

Father (Captain) John L. O'Donnell was in the thick of the fighting from September 26 onward. His Distinguished Service Medal citation read:

> As regimental chaplain, he was ceaseless in his efforts to better the welfare of the men and during the period of operations he accompanied the attacking waves in every action in every action in which the regiment took part. Exposing himself to artillery and machine gun fire to care personally for the wounded, organizing parties of stretcher-bearers, going without thought of personal danger whenever he was needed, he set an example of courage and heroism, appreciably raising the morale of those with whom and for whom he worked.

The 33rd Division was key to the success of the Meuse-Argonne Offensive. The American front was thirty miles wide and the 33rd Division was on the extreme right. The 33rd was to advance north and swing east onto Meuse River. The entire American line off to the west pivoted on the 33rd Division.

After the success of September 26, the 33rd Division held the west bank of the Meuse River from Sept 27 through Oct

7, 1918. The Germans maintained continuous artillery fire from the east bank throughout that period. There was considerable gas shelling with yellow cross (mustard) gas and chloropicrin. The latter induced vomiting, causing the soldiers to remove their face masks and then breathe the mustard gas into their lungs. On the night of September 29-30 in the Bois de Forges area, the Germans lobbed an estimated fifty 105mm yellow cross shells on the 2nd battalion/132nd Regiment, which included Captain Mallon's Company E. The doughboys did not evacuate the area and sustained many gas casualties. The eerie whistling and screeching of an incoming shell will cause most men to crouch or hit the dirt if they have time. Perhaps they can manage to protect themselves from shrapnel or even simply the resulting shower of dirt and wood from hitting their face. If the explosion hit, there was nothing anyone could do to escape injury or death.

On October 1, a high explosive shell badly wounded Captain Mallon in his right thigh. He lay unconscious. Two men evacuated him from the field. Triage treated Mallon and sent him to the rear. He was at Base Hospital No. 35 in Mars, France. While in the hospital, Mallon had time to himself. He wrote a friend in the Twin Cities in October.

> I got bumped [wounded] on October 1 and I may be laid up quite a while. I thought it didn't amount to much at first … A high explosive shell hit beside me and a piece tore through my leg, about halfway between the knee and hip. No bones were broken but the flesh is busted up pretty much … I hope it gets well in time for me to get back and get a few more cracks at Fritz. We had been giving them hell before I got hit …

The above letter appeared in the *Minneapolis Morning Tribune* under the eye-catching headline, "Ten Yanks Clean Up Whole Hun Battery."

Capt. George Mallon with Sgt. Sydney Gumpertz (left & third from left) with comrades of the 132nd Infantry Regiment, 33rd Division in France
courtesy of George Mallon

Mallon did not return to the war. Instead, he remained at Base Hospital No. 35 in Mars, France to the end of the war and beyond. The hospital did not release him until January 14, 1919. As he later said, "October 1st, to be exact – I was severely wounded by a high explosive shell. That ended the war for me."

The State of Minnesota established a Minnesota War Records Commission after the war. The commission sent a four-page questionnaire to every veteran to serve as a permanent "Military Service Record". In response to "Resumed former activities in civil life?", Mallon answered, No." In response to "If not ... why?", Mallon answered, "Did not care to follow former occupation as injured limb was not as strong as formerly." For his wound, Mallon received $150 per month for the rest of his life.

CHAPTER NINE
MEDAL OF HONOR
AND
ONE OF PERSHING'S 100 HEROES

In early February, Captain George Mallon and First Sergeant Sydney Gumpertz received orders to report to AEF Headquarters at Chaumont, France.

Mallon later noted, "I believe I was the most surprised man in the A.E.F. when I received notice to report to General Pershing's headquarters to receive the Medal of Honor." He took some comfort in knowing that his good friend and comrade also was going to be awarded.

On February 9, 1919 at Chaumont, General Pershing presented the Medal of Honor to seventeen men. They were, from (General Pershing's) left to right:

    Captain Edward C. Allsworth, 60th Infantry
    Captain George H. Mallon, 132nd Infantry
    Captain George C. McMurty, 308th Infantry
    Lieutenant Samuel Woodfill, 60th Infantry
    Lieutenant Harold A. Furlong, 353rd Infantry
    2nd Lieutenant Donald M. Call, Tank Corps
    Sergeant Johannes S. Anderson, 132nd Infantry
    Sergeant Sydney Gumpertz, 132nd Infantry
    Sergeant Willie Sandlin, 132nd Infantry
    Sergeant Archie A. Peck, 307th Infantry
    Sergeant Harold I. Johnson, 356th Infantry
    Corporal Frank J. Bart, 9th Infantry
    Corporal Jesse N. Funk, 354th Infantry
    Corporal Berger Lohman, 132nd Infantry
    Private Charles Barger, 354th Infantry
    Private Thomas O. Biebour, 167th Infantry
    Private Clayton Slack, 124th MG Battalion

Five of those being honored were men of the 132nd Infantry. Of course, Mallon and Gumpertz served in Company E, 132nd Infantry. Willie Sandlin of Kentucky was with Company A. Like Mallon and Gumpertz, he also received his Medal of Honor for his gallantry on September 26, 1918, the opening day of the Argonne Offensive. Johannes Anderson of Chicago (born in Finland) served in Company B. His decoration was for his valor on October 8, 1918. Berger Lohman, also of Chicago (born in Norway), served in Company H and was honored for his courage on October 9, 1918. Anderson and Lohman's actions were from the crossing of the Meuse River, a week after Mallon was wounded and evacuated from the field.

Captain Mallon's Medal of Honor citation read:

> Becoming separated from the balance of his company because of a fog, Captain Mallon with nine soldiers pushed forward and attacked nine active hostile machine-guns, capturing all of them without the loss of a single man. Continuing on through the woods, he led his men in attacking a battery of four 155-millimeter howitzers, which were in action, rushing the position and capturing the battery and its crew. In this encounter, Captain Mallon personally attacked one of the enemy with his fists. Later, when the party came upon two more machine-guns, this officer sent men to the flanks while he rushed forward directly in the face of fire and silenced the guns, being the first one of the party to reach the nest. The exceptional gallantry and determination displayed by Captain Mallon resulted in the capture of one hundred prisoners, eleven machine-guns, four 155-millimeter howitzers and one antiaircraft gun.

Victory Loan Poster memorializing Captain Mallon
produced in 1919 for the Victory Liberty Loan Committee
reproduced in The *Nonpartisan Leader*, June 14, 1920

On the day following the ceremony, one limousine with four stars and two limousines with three stars transported the seventeen Medal of Honor recipients to General Pershing's chateau, Val des Escaliers, for a casual luncheon. There was much mixing between the ranks with small chat and jokes over drinks. The generals saluted the captains and lieutenants, as well as the sergeants, corporals and privates. It was a special experience. General John J. Pershing knew who Captain George H. Mallon was. The incident received considerable coverage, including a front-page article on the February 14, 1919 issue of *Stars and Stripes*. The article noted, "Never before in the history of the American Army has such an array of stars and bars and just plain O.D. mixed in common in the dining room. And the affair was not a stiff, formal one, either ... talked and exchanged jokes." *Stars and Stripes* later printed the full citation for all forty-seven Medal of Honor recipients on page 5 of its March 14, 1919 issue.

Just weeks after the Medal of Honor ceremony at Chaumont, Secretary of War Newton Baker asked General John J. Pershing for stories about the war that could be used to promote the purchase of bonds to help the federal government finance the war effort. These were not just bonds, but Liberty Bonds. Baker led a huge effort to propagandize the war to pay for it. The Liberty Bonds paid at slightly lower interest rates than a typical savings account in the bank. There were four Liberty Loan drives during the war. The fifth Liberty Loan issue – the Victory Liberty Loan – took place after the war on April 21, 1919.

Pershing and his AEF staff put out a request to the division headquarters in France. Those division staff groups, which moved quickly to respond, ended up with more of their decorated soldiers used in the promotion of the Victory Liberty Loan. Timing was everything. In the end, there were a multitude of obvious omissions. Among others, Alvin York, the most famous hero of the war, was not included. The 33rd Division apparently was timely. Among those included in promoting the Victory Liberty Bond was Captain George H. Mallon. In addition, the honorific as "One of Pershing's Hundred Heroes" remained with Mallon for the rest of his life. In a way, that seemed to become over time almost more important than "Medal of Honor recipient".

A newspaper noted, "The Victory Liberty Loan campaign will be the greatest ever undertaken in the history of advertising." The committee prepared more than one hundred full-page advertisements and smaller pieces of copy and distributed them to every newspaper in the country. A portion of the money raised was used to pay off the cost of the advertising effort. A member of the committee was quoted as saying, "If this kind of copy doesn't stir 'em up and make them over-subscribe the loan, then I don't know what will!"

The graphic of Captain Mallon depicted him throwing his famous punch to the jaw of the German officer. That forever cemented his image as a tough guy. Back in the U.S.A in Washington, D.C. on March 28, journalist A. E. Geldhof celebrated one of America's newest heroes, writing: "Give Captain George Mallon a handful of men and he'd attack the whole German army – and he'd give it a beating too." The honorifics built one upon the other.

The government used Captain Mallon as "the chief feature of exploitation in the Victory Loan drive in Kansas City, Kansas." The Chamber of Commerce sent thousands of an immense poster "… out to every section of the city and county." The newspaper described the Mallon poster as "a most effective bit of heroic art." There was speculation that the decorated officer would "… have a far more difficult time escaping the honors of political office than he ever encountered in his fight with the blood-thirsty Huns."

Captain Mallon was among 96,000 American soldiers wounded in the forty-seven day Meuse-Argonne Offensive, which ended the World War. Over 26,000 Americans lost their lives in the battle. Of Mallon's Company E alone, 42 men were killed in action, another 79 were wounded. Three others died and ten were captured.

The *Minneapolis Journal* featured a huge headline: THE

WAR IS OVER. The people of Minneapolis rose from their beds to march, sing and cheer. Church bells competed with sirens and cars honking their horns. November 11th was a day to be remembered throughout ever city, large and small, in the United States.

George Mallon wrote Clos Egan, Business Agent for the Steamfitters Union, from Paris on January 19, 1919. "I started for Germany and I figured I'd get there and now I'm on my way." He spent the rest of occupation with the 33rd Division.

Both Minneapolis, Minnesota and Kansas City, Kansas claimed Captain George Mallon as their own. The Kansas City press proclaimed, "MALLON ON WAY. One of War's Greatest Heroes Is Homeward Bound. HE IS A K. C. K. PRODUCT." The article noted the Kansas City address: "Mrs. Mallon's home is at 1931 Hallock Avenue." Effie Mallon was living with her mother at the time.

Secretary of the Navy Josephus Daniels sailed home across the ocean with men of the 33rd Division. He told a reporter:

> A finer looking body of men I have not seen. Having had an eight-day shipmate association with the 33d, I came to admire them because they have true American stuff and pep and character. They are the type of men who cannot be defeated and who are coming home to civic duties which they will perform with as much patriotism as they served their country during the war.

The 132nd Regiment/33rd Division landed at Hoboken, New Jersey on May 17th, 1919. A contingency of Illinois dignitaries, included Governor Lowden and Senator Medill McCormick, greeted the doughboys as they went ashore. Major General George Bell extolled the merits of the men of his division. "Hamel, I consider, was one of the most important battles of the war because the spirit shown gave our armies an insight into the character of the men we were fighting with."

The *Chicago Sunday Tribune* of May 25, 1919 pumped up the hometown crowd, declaring in large type on its page one headline: ALL OF 33D BACK IN THE U.S. The sub headline read: FIRST UNITS WILL BE HERE ON TUESDAY. Further in the same issue, an editorial titled "America and the Treaty" questioned whether the Treaty of Versailles "... is the fertile seed plot of future wars."

On Tuesday morning, The *Chicago Daily Tribune* headline read: FIRST PRAIRIE MEN HOME. An elaborate illustration on the front page showed crowds of people hanging from every window of every story of a building with flags waving.

Major General Leonard Wood was commander in chief of the military demonstration. The parade began at 11:00 a.m. The line of men marched north on Michigan Avenue to Randolph, then west on Randolph to State Street, then down State to Adams Street and in this way crisscrossed through the canyons of downtown. Major General George Bell, followed immediately by Division headquarters, led the parade. Brigadier General E. L. King, accompanied by his 65th Infantry Brigade headquarters, was next. Behind King was Colonel Abel Davis and his 132nd Infantry Regiment, which included Captain George Mallon and the men of Company E. Many other units followed the 132nd.

The Prairie Division marched in a triumphant welcome home parade down through the heart of Chicago's Loop. Wednesday's *Tribune* reported the event:

> Chicago yelled its head off. Chicago shouted to colonels and majors and captains and privates and second lieutenants and non-coms and orderlies. Chicago whistled and roared and cheered and rang bells and beat gongs and blew horns and threw flowers ... cheering and crying all at once ... The sun was on the lake. The sun was shining on Michigan Avenue. Chicago's imposing skyline was before them ... palisades that were filled with caves

and in each cave men and women with flags and white arms that waved welcome. Michigan Avenue! As far as the eyes could reach were people, along both sides of the asphalt, on the tops of buildings … Everywhere was color; everywhere was noise.

You could see them coming now, soldiers in platoon formation, following the horses … Maj. Gen. George Bell Jr. and Lt. Col. Simpson, his chief of staff, in the lead. Then came the officers of the division staff, the headquarters troop, and then the doughboys, the 132nd Regiment with Col. Abel Davis walking at its head. The band was playing "Illinois" … Officers and men looked right and left, smiled to the cheering thousands, waved hands, passed on.

At the end of the parade in Grant Park, the thousands of returning soldiers made their way to the various assigned hotels for the banquets, which awaited them. Captain Mallon and the thousand-plus officers and enlisted men of the 132nd Regiment made their way to Hotel Sherman at the corner of Randolph and Clark Streets (present day site of the James R. Thompson Center). The magnificent hotel, constructed in 1910, was the largest hotel in Chicago. The soldiers made their way to the hotels with their arms around loved ones. "Their faces were wet with sweat and multi-colored chunks of confetti clung to heads and hands and necks." Passerbys shook the hands of the soldiers and slapped them on their backs. Many of the men of the 33rd Division were from Chicago and nearby towns. Captain George Mallon and only fourteen of the two-hundred-plus men in Company E were from Minnesota. There was no one to greet them personally. Meanwhile, all around them, loved ones reunited.

Stephen D. Chicoine

Left to right: Captains Edward C. Allworth, George H. Mallon, George G. McMurtry, 1st Lieutenants Samuel Woodfill, Harold A. Furlong, 2d Lieutenant Donald M. Call, 1st Sergeants Johannes S. Anderson, Sydney G. Gumpertz, Sergeants Willie Sandlin, Archie A. Peck, Harold I. Johnston, Corporals Frank J. Bart, Jesse N. Funk, Berger H. Loman, Private 1st Class Charles D. Barger, Privates Thomas C. Neibaur, Clayton K. Slack

Officers and Men Being Presented with Medal of Honor at Chaumont, February 9, 1919

Medal of Honor recipients (Mallon is second from left)
Chaumont, France, February 9, 1919
from American Armies and Battlefields in Europe

Most of the eight men of Company E, who were with Captain Mallon when they took the four big howitzers on September 26, 1918, were from Chicago. Harry A. Dawson of Chicago was a corporal at the time. He discharged from the service as a sergeant. Two of Mallon's band were Chicago residents, who were not yet citizens when they were inducted. Anton Churas was born in Lithuania and Sam Salpietro was born in Sicily. Salpietro discharged with the rank of corporal. Vivian C. Badger, a railroad clerk in Chicago, was in the Illinois National Guard before the war. He discharged from the Army as a sergeant. Private Allen W. Bridges was from Chicago. Private Albert F. Wittman of Chicago later died in action on October 9, when the regiment crossed the Meuse River. Private William Kelly was a Nebraska native, working as a farm laborer in Iowa before the war. He also died in action on October 9. Private Gustave A. Schill, was from Pennsylvania.

The return of the Prairie Division was one of the greatest events in Chicago history. The doughboys of the 33rd Division settled into Camp Grant in Rockford, IL, ninety miles northwest of Chicago. There must have been considerable uneasiness as over 1,000 doughboys died of influenza at Camp Grant in Sept-Oct 1918. No one wanted to die after surviving the war and returning home.

CHAPTER TEN
WORLD WAR VETERANS

The Americans, who went off to fight in France in the World War, returned home determined to actively participate in the political process. To that end, a group, which called itself the World War Veterans, organized in France on November 20, 1918. Another veteran organization, the American Legion, formed in France, four months later, on March 16, 1919. Among the founders of the American Legion was Colonel Theodore Roosevelt, Jr. It was the opinion of World War Veterans that the American Legion was aligned with aristocracy and big business. World War Veterans was pro-labor and would not accept anyone, who worked as a scab during a strike, into the organization. Both were determined to see things their way and the two would clash for the next several years. Many veterans felt that a government, which claimed the right to conscript for war, had a responsibility to see that the experience did not financially ruin those who returned.

The World War Veterans led the labor division in the Minneapolis Memorial Day parade in May 1919. They chose Captain George Mallon to lead them in the parade. However, Mallon was still at Camp Grant in Illinois and unable to attend. The Army did not discharge him until June 20, 1919 at which time George Mallon returned to the outside, as he and his comrades referred to civilian life.

George Mallon returned home to Minneapolis to much acclaim. The national press told and re-told stories of his exploits in France and often referred to him as being from Kansas City, Kansas. From that point onward, both states rightfully claimed the war hero as their own.

The *Evening Star* of Washington, D.C. reported:

... Capt. Mallon said he knocked down the enemy

battery commander with his fist when the German drew his revolver to shoot him. Asked why he did not shoot, Capt. Mallon replied: "I couldn't kill a man in cold blood." At the same time a German private behind one of the gun carriages tried to kill Capt. Mallon and was shot down by an American.

That interview revealed insight into Mallon's use of his powerful fists; i.e., their use allowed him to accomplish his mission without needlessly taking any more lives than was necessary.
George Mallon's confidence was matched by his profound humility. He recognized that the immortalizing of heroes was part of the collective experience of war and that it served the purposes of the military. Mallon did not embellish the truth, yet he could see the story of his exploits growing and taking on an element of legend. He did his best to turn the story to he and his men, versus that of the lone, heroic warrior. Any experienced veteran immediately noticed that the chaos of combat, the loss of life and the horrific sights and smells were missing in the legend.

Days after returning to Minneapolis, Mallon met with his friend, former Minneapolis mayor Thomas Van Lear. Van Lear, like Mallon, was a regular Army veteran of the Spanish American War. More importantly, both Lear and Mallon were trade unionists. They shared the radical view that labor had a right to strike and that law enforcement had no right serving as the tool of the owners by brutally suppressing strikes.
Van Lear was mayor through 1917 and 1918. The Socialist Party expelled Mayor Van Lear in 1918 when he refused to support the resolution for active militant opposition to U.S. involvement in the war in Europe. Republicans mounted an aggressive smear campaign against Van Lear during his bid for re-election. Voting took place on November 5, 1918 while the World War was still on. The opposition claimed that Van Lear hoped for a German victory. The tactic was successful. Van Lear lost by 1,300 votes.

Thomas Van Lear subsequently became involved with the Nonpartisan League. His son, Ralph, was one of the organizers of the Minnesota section of the World War Veterans. This connection, no doubt, led to Mallon's affiliation with the World War Veterans. The two organizations had similar progressive objectives.

The Nonpartisan League began as a movement in North Dakota in 1915 as a reaction to Minneapolis and Chicago millers, bankers and hedgers, who benefitted so greatly from the hard work of farmers raising grain. Founder Arthur Townley was a failed flax farmer. The Nonpartisan League's objective was simply "... to take the government out of the hands of Special Privilege and restore it to the people." Special interests did their best to present that position as extremely radical and to label any supporters as "Red" or "Bolshevik".

Townley sought to circumvent the two-party system, which political bosses controlled and was part of the system of exploitation. His plan was simple; i.e., to concentrate the votes of its League members in one of the two party primaries and tip the balance. That could be to support a specific party candidate who shared nearly similar views, but mostly involved placing a League member into the primary as a candidate. The name Nonpartisan reflected the fact that either party could be a vehicle. Townley's organization was a movement, not a political party. If the League lost a primary, Townley then ran his people in the general election as independent third party candidates. The Nonpartisan League co-opted the Republican Primary in North Dakota in 1915 and gained control of the state government. The League established a similar organization in Minnesota in early 1918, much to the distress of the political bosses and big business.

The July 18, 1919 issue of the *Minneapolis Labor Review* reported that the recently returned hero Captain Mallon was the unanimous choice of the delegates to become the Business Agent for the Business Trades Council. The crowd called for a speech and Captain Mallon gave what the *Review* reported as a "modest talk ... there was nothing of bravado about it. There was nothing to suggest that he had done anything out of the ordinary...". Mallon showed his leadership abilities when he said:

> Why, if you men hadn't stood behind us we couldn't have done anything on the other side. We couldn't have even gotten started. Everything we were able to do, we owe to you workers standing back of us. The war proved what organization means and now we must have more perfect organization than we have ever had before,

George Mallon was a man of action and an effective leader. By August 1, Mallon effected the settlement of a strike and organized the union membership of the glaziers.

There was a tidal wave of labor strikes in 1919, at least in part, related to the post-war economic depression. Reactionaries proliferated in the government. Foremost of these were Attorney General A. Mitchell Palmer and J. Edgar Hoover, whom Palmer appointed as head of the General Intelligence Division of the Justice Department. Hoover ran his organization as an anti-radical, anti-subversive spearhead of the government. Palmer, Hoover and many others perceived the organized labor movement as part of complex sinister plan to overthrow the government. It was the beginning of the Red Scare. One of the basic founding elements of the American Legion was to stay out of politics. However, the battle lines were soon drawn between labor "radicals" and elements of the American Legion as strike-breakers. The war of words carried into fights in the streets.

Page 2 of the August 22, 1919 issue of *Minneapolis Labor Review* included an article headlined, "Why Returned Service Men Will Organize" with sub-headings, "Good Enough to do Own Fighting, Good Enough to do Own Talking" and "World War Veterans Not Controlled by Officers":

> World War Veterans ... is not dominated by high

officers nor any outside interests ... We oppose the exploitation of our patriotism and loyalty by any group of people for their selfish interests. We do not want to be used as scabs in case of a labor strike, nor do we want to be used as a lever to force down the wages of other citizens. Nor do we intend to starve or beg ... there is ample opportunity for all to work if government will only let down the bars of monopoly and privilege ... We are opposed to universal military training ... The World War Veterans at the present is the only ex-service men's organization which has the indorsement of the Minnesota State Federation of Labor.

By Labor Day 1919, George Mallon was a member and active spokesman for the World War Veterans. Twenty-five thousand marched and Mallon served as Grand Marshall of the Minneapolis Labor Day parade. He addressed the crowd:

I am speaking on behalf of the organization of World War Veterans, an organization of returned service men which has the only indorsement of organized labor.

All a soldier can give is his life for his flag and I know you would do it. But it is not only in war, but in peace, that you must be watchful for the government. You have got to do it. The profiteers said they were back of you. They were to the extent that they made 17,000 millionaires while you were away. They are still back of you. They want you too patriotic to ask for that bonus and to help them keep the millions they made out of the war.

I'm for the bonus. Most of you came back broke. All my company were broke ... Don't let the political grafters lead you who say they can't give you more wages. Tell them you want every dollar you earn every day you work. Most of the men who won the war were laborers ... You got $30 a month and came back and got nothing and they say that was too good for you.

I belong to the World War Veterans and hope to see you all there.

Captain Mallon's remarks as to the profiteers were bold, if not dangerous. He certainly was aware of that, but George Mallon was not one to cower in safety. The Sedition Act of 1918 allowed the federal government considerable leeway to infringe on freedom of speech. As *Minneapolis Labor Review* (Sept. 5) noted, "Even if you have been a soldier unless you agree to the bidding of big business, you are a Bolsheviki."

War profiteering was a major theme for the newly-formed World War Veterans, as it had been for the Socialists and the Nonpartisan League, long before Captain Mallon returned home from the war in Europe. The Commission of Public Safety, the infamous Minnesota watchdog agency, deemed the Nonpartisan League as failing the American loyalty test in 1917. The U.S. government sentenced Socialist Rose Pastor Stokes to ten years in prison in 1918 after she accused the government of being allied with profiteers. Arthur Townley of the Nonpartisan League was arrested in Minnesota during the war for criticizing the profiteering aspect of the war. What Townley actually did was call for the conscription of wealth, when the nation began conscripting young men for the war.

The Sedition Act of 1918 was inoperative when Mallon spoke, given that the war was over. That did not mean he was safe from prosecution. At the time of Mallon's Labor Day talk, Attorney General A. Mitchell Palmer was in the midst of campaigning for the Democratic Party's nomination of president for 1920. Palmer urged the passing of a peacetime Sedition Act so there would be no criticism of the government. Palmer later told a primary crowd: "I am myself an American and I love to preach my doctrine before undiluted one hundred percent Americans, because my platform is, in a word, undiluted Americanism and undying loyalty to the republic."

Thomas Van Lear also addressed the vast Labor Day crowd in Minneapolis.

> What are they going to say about the men wounded in Flanders and France who are members of the World War Veterans? The same as about me. They said, 'Oh, Van Lear was loyal in the Spanish-American War, but since that time he joined the Socialist Party and is disloyal.' They are going to say the same thing about Captain Mallon and the World War Veterans. The newspapers devote columns to the sons of influential men, but they give practically no space to Captain Mallon because they don't like his politics and because he is a leader of labor and big business is not behind him.

On September 17, 1919, just two weeks after his Labor Day address, Mallon spoke to workers at noon at the Minneapolis Steel & Machinery Company. He attracted a much larger crowd than any previous speaker. Captain Mallon began by asking those in the crowd, who had been in the service, to raise their hands. Many responded.

Mallon:	What did we fight for over across the sea?
Crowd:	Democracy!
Mallon:	Are you getting it in this plant?
Crowd:	(little to no response, negative rumbling)
Mallon:	You boys ought to stick together in the union here, just as we stuck together across the ocean.

Mallon had the crowd in his hands. He also had the attention of the employer and of Citizens Alliance.

On September 19, 1919, the Minnesota State Legislature approved a bonus for soldiers, who served in the World War. *Minneapolis Labor Review* on September 26 noted:

> The State headquarters of the World War Veterans is receiving many messages of congratulations since the passage of the Bonus Bill. These messages come from different parts of the state and the boys seem to be highly pleased with the big fight the World War Veterans put up to have that bill passed.

On Friday night, September 26, 1919, Captain Mallon and Tom Van Lear addressed more than 2,000 people at St. Paul Auditorium. The event was under the auspices of the Minneapolis Council of Railroad Workers. Both men were in an aggressive mood and determined to respond to recent attacks. Big business was openly contributing sizeable amounts of cash in slush funds "to combat Socialism." One pamphlet circulating through the Twin Cities was titled, Russia or America? Choose. The pamphlet slandered the workers, the labor movement and the World War Veterans organization. At the same time, the Steel Strike of 1919 was underway, having commenced on September 21. Men and women were dying on the strike line.

Mallon spoke forcefully and the audience cheered wildly.

> I didn't intend to join any ex-service men's organization, but when I read the World War Veteran's pamphlet and realized that nobody but soldiers were going to handle that organization, I joined... The World War Veterans realize who won this war and nobody else. A man may not belong to organized labor, that is his misfortune, but no

one can deny the workers are being exploited … I came to the conclusion that what the exploiters want is to make the solder so patriotic he will help them hang on to their profiteering money. I can't understand how these 17,000 millionaires got this dough if they were fighting so much. There was a time when they would have done anything for the soldier when they didn't know how the war was going. All through the East they are trying to organize soldiers as strike breakers and there is a returned soldiers' organization offering 5,000 men in uniform to guard the scabs or work in the mills.

Mallon was referring to the American Legion, but he specifically did not name the organization in his speech.

I have thought lots of my men in France. We lost 47 killed and 82 wounded, among them two lieutenants. I have thought lots also of the men who sleep in France, but I am not so sure they are better off than the men forced to scab. I WOULD RATHER BE BURIED IN FRANCE TODAY THAN BE FORCED TO SCAB. It is not right for the Government to let those men force the boys to scab. The men came out of the service without any money … So you must know the boys came back broke. They asked for $25 bonus and were turned down, but the World War Veterans kept up the fight for the bonus and finally it was obtained.

They talk about Americanism. I am an American first, last and all the time. If this Government fails, I don't know any place I want to go. But I don't want any little clique of profiteers telling us what Americanism is. The slogan of the World War Veterans is enforcement of the Constitution as it is written. I haven't seen much in the papers about the World War Veterans in the papers. Sometimes we are called Bolsheviks. And when they call us that we want to tell them that it must have been the Socialists and Bolsheviks that won this war. You can't fool old American buck private very much and he knows what they are trying to do. It wasn't hard to be an officer. I was one and I know ... We are all against the war. I am against war myself. The conscientious objectors did their duty. I know because I signed their discharges myself. They have a right to their belief ... I'd like to see every soldier in the World War Veterans. It is your duty to yourselves and your comrades to join.

Minneapolis Labor Review noted, "Mallon made a splendid address ... was greeted with a wonderful ovation when he finished speaking."

Tom Van Lear's address to the crowd received considerable attention from the opposition and the mainstream press. The *Review* quoted Van Lear:

> Most intelligent working people have come to realize that the promises of reconstruction, the slobber during the war about the employers and the workers sitting down after the war was just bunk. The capitalists said that because they didn't want to be interfered with while they were making fortunes. They said Homestead and Ludlow were behind us ...

Workers struck at Andrew Carnegie's steel works at Homestead, Pennsylvania in June-July 1892 and at John D. Rockefeller Jr.'s coal mines in Ludlow, Colorado in April 1914. Both involved Pinkerton agents, state militia and national guard in open warfare and violence with striking workers and their families with the result of considerable loss of life. Van Lear went on to say:

> The war was hardly ended before they commenced to organize to fight labor ... The American Committee is also the judge of everyone's patriotism. Whether you were the highest decorated or left some of your limbs in France, this committee decides your patriotism on how they can use you. Their sacrifices are not proof enough that they are patriotic. They have got to join the American Legion. If they join the World War Veterans, they are not patriotic. Here is Captain Mallon the highest decorated man in Minnesota, they have no use for him because he opposed them and they attack the organization of which he is a member. His offense is that he opposes them in their exploitation. If you do that they question your loyalty ... They want them to join an organization to be used to protect scabs.

Van Lear proceeded to attack the American Legion for being a tool of Big Business and protecting scab workers.

> Your organizations of labor will never be safe as long as corporations have the power of the gun behind the table. But when the guns set on your side of the table, you will be able to make such agreements as you never made before. The papers said Van Lear wants to get the workers behind the guns to shoot and hang the capitalists. That's because they had had bad consciences. They have done it to us so often they think we would want to retaliate but that is not true. We want them in our possession so they can't use them to kill us.

It was rare for the *Minneapolis Morning Tribune* to provide any coverage of labor events to its large readership. However, in the aftermath of the Railroad Workers event, The *Tribune* lashed out. The headline of the October 1 issue read:

> Van Lear Flays Legionaires in St. Paul Speech
> Asserts Front Line Soldiers Not in Sympathy with Organization
> Accuses "Big Business" of Being Behind Newly Formed Order

The *Tribune* referred to the address as "A vicious attack" and made a point of not attacking Captain Mallon. The paper only shared his explanation of the World War Veterans on conscientious objectors. There was no other mention at all of Mallon's address, including some of his more pointed remarks, which were the thrust of his address. As with others, *Tribune* management either respected the man, despite their issues with his stands, or preferred to keep his politics hidden from the public. The *Tribune* devoted two full columns to attacking Van Lear. The *Tribune* also reported that Harrison Fuller, state chairman for the American Legion, would be issuing a formal statement "... giving some facts in connection with the World War Veterans." The paper quoted Fuller tossing out the "anti-Americanism" label. There was also the matter of a debate proposed by Lester Barlow, head of the World War Veterans.

Two days later, on October 3, the weekly *Minneapolis Labor Review* responded with the headline:

> St. Paul Meeting Goads Kept News Papers to Fury
> Captain Mallon Says Rather Be Buried in France Than Scab on Job

The *Review* gave considerable coverage to both Captain Mallon's speech and that of Van Lear and commented, "Tom Van Lear did a brave thing."

Minneapolis Morning Tribune of the same date, October 3, on page twelve published a full-page attack by the American

Legion, asking "Has Van Lear Right to Slur Men Who Fought." There was no mention of Mallon.

Harrison Fuller's formal response appeared in the *Tribune* on October 6, 1919.

> Because I fail utterly to see any "issues" as between the World War Veterans and the American Legion, I am compelled to decline Mr. Barlow's challenge to an open debate on the merits of the two organizations ... I should like to assure Mr. Barlow that at no time has the integrity, loyalty or sincerity of the World War Veterans been questioned by any official of the American Legion. As long as the World War Veterans speak as men and soldiers, they speak a common language with the American Legion.

Of course, there were major differences between the two organizations. Fuller went on in his letter in the *Tribune* to address a most interesting aspect of the entire conflict, the common respect by members of both organizations for Captain George H. Mallon.

> I cannot for a moment believe that Mr. Van Lear speaks for the World War Veterans or that they would tolerate for one moment the counsel or friendship of such a man as he. I am more certain of this because Mr. George H. Mallon, representing the World War Veterans, who spoke from the same platform, made not the slightest reference to the American Legion, avoiding as carefully as we have done the casting of any slurs on a kindred organization.

Fuller and Mallon attempted to maintain a delicate balance. Mallon had many good friends in the Legion, even if politics was not something upon which they agreed.

New Ulm Review of October 15, 1919 did its best to escalate matters in its distribution area, reprinting an article from *McLeod County Republic* titled "Decorated Hero Slams Legion":

> In a speech in the St. Paul Auditorium ... by Capt. Geo. Mallon, the American Legion was charged to have been organized in the interest of the reactionary politicians and that the idea in starting it was to make of the organization an efficient tool of big business ... it will be hard for the old gang to accuse him of being disloyal so, instead, the daily papers almost forgot the fact that he spoke at all and played up Thos. Van Lear's speech, which was tame in comparison to that of the returned soldier. The hypocrite press may have succeeded in making a few people think that Mr. Van Lear was disloyal, but they will have a hard time making them believe that Capt. Mallon is anything but a patriot.

The conflict between the World War Veterans and the American Legion became more and more apparent by the day. Lester Barlow of the World War Veterans reserved an outdoor venue in Waterville, Minnesota to speak to farmers in the area. The event was to take place on Monday evening, October 13, 1919. *Minneapolis Labor Review* of October 17 reported that members of the Legion post being formed in Waterville solicited local businessmen to sign a petition that Barlow not be allowed to speak. The *Review* noted, "Barlow had booked the meeting at the request of men who had been in the service who are residents of that place. It was threatened that the meeting would cause a riot." Farmers, upon hearing of the petition, responded by planning massive participation. For the good of all, the weather turned bad and there was no physical clash in Waterville. The fight was not over. The local Legion posted a notice in the Waterville paper, asserting that the Legion was the real organization and that all loyal men belonged to the Legion. The *Review* printed Lester Barlow's response, which included:

> The most decorated returned soldier in Minnesota is Captain Mallon. He is not a member of the American Legion, but is a very active member of the World War Veterans in Minnesota.

One of several major differences between the World War Veterans and the American Legion was the issue of universal military training. The concept was that all qualified citizens would serve in active and reserve duty to allow quicker mobilization of an army in time of need. The American Legion supported universal military training. The World War Veterans did not. *Minneapolis Labor Review* published a piece by George Mallon on this subject on November 14, 1919. The *Review* included large headlines and a photo of Captain Mallon in uniform:

> I am opposed to Universal Military Training as I see what it has done for the people of nations that accepted it. Militarism is one of the things I fought against in Europe and I had hoped the world would be rid of ...

Mallon followed that clear statement by connecting it to his ongoing theme of war profiteering:

> There is, I am sure, a class of people in this country of ours, who would plunge the nation into war at any time for the "blood money" they would make out of it, incurring no risk whatsoever to themselves as the "common people" carry the burden of all wars. I do believe in a moderate "standing army" and a good navy, but am in terror of militarism ... What we need in America is men in office who will carry out the will of the people

and not be accountable to the few "profiteers".

The same page of the *Minneapolis Labor Review* included an article in which Lester Barlow of the WWV wrote of his disappointment with the American Legion. The headline read: "Legion Fails to Criticize Profiteers." Barlow attended the first national convention of the American Legion, which was held in Minneapolis from November 10-12. He appeared before the Resolutions Committee to ask, "Why don't you get after all the anarchists, the grafter and the profiteers, as well as the alien and the slacker?" The *Review* reported: "This aroused the wrath of the Committee and the guest was invited to leave in no uncertain terms."

The issue of Universal Military Training (UMT) was the subject of much national debate in 1919. General John J. Pershing testified for three days before the Joint Senate and House Military Affairs Committee in October. He urged the nation to adopt Universal Military Training, calling for eleven months training for every able-bodied young American, followed by service in the reserve for four years.

George Mallon was in direct opposition to Pershing, a man who was strong-willed and demanded loyalty. Despite common reference to him as Captain Mallon, George Mallon no longer answered to Pershing or any other superior officer. He was George Mallon, labor man and champion of the working class. *St. Louis Post-Dispatch* in its December 14 issue pointed out that Captain Mallon made no distinction between military training and militarism. Mallon maintained from experience that men could be turned into soldiers with six months training.

The politics of men, such as George Mallon, did not go unnoticed. The Federal Government, determined to protect the status quo at any cost, directed United States Military Intelligence (USMI) to spy on American citizens. The government charged USMI with "the surveillance of all organizations or elements hostile to or potentially hostile to the government of this country or who seek to overthrow the government by violence".

Authorities believed that included anyone who spoke for change for the common workingman. That included George H. Mallon of the World War Veterans.

On October 24, 1919, Lt. Col. John B. Reynolds of the Headquarters Central District in Indianapolis sent a memo to Col. Mathew C. Smith of the Service and Information Branch, Office of The Assistant to the Secretary of War. Their responsibility extended to anything and everything concerning veterans. Reynolds wrote:

> Attention is called to the article regarding "Captain" George H. Mallon of the World War Veterans in the enclosed copy of the Minneapolis "Labor Review". If this alleged Captain has any status in the Reserve, his activities along radical lines would seem to warrant an investigation to determine whether or not he is entitled to a Reserve Commission.

The only article in the *Review*, which mentioned Mallon in October, concerned the speech, which he made at St. Paul Auditorium to the Minneapolis Council of Railroad Workers on September 26. While the American Legion and *Minneapolis Morning Tribune* chose to target Van Lear rather than Captain Mallon, United States Military Intelligence did not hesitate. Lt. Col. Reynolds clearly had no idea that the provocative speaker was a Medal of Honor recipient.

The situation in Minnesota. On November 17, 1919, former U.S. Congressman Ernest Lundeen took the stage in Ortonville, Minnesota to speak out against the League of Nations. He was noted for being one of fifty U.S. Congressmen to vote against the United States declaration of war against Germany, which lead to U.S. entry into the Great War. His term as a congressman ended just months earlier in March 1919. A mob of American Legionnaires forcibly removed Lundeen from the stage and locked him in a refrigerator car on a train bound out of town. The local sheriff was an accomplice to the act. *Minneapolis Labor Review* headlined, "What Will Governor Do in Lundeen Mobbing?" and referred to the incident as "disgraceful and despicable." The World War Veterans used their slogan. "The Enforcement of the Constitution As It Is Written," to call for justice.

The adjacent column of the same issue of *Minneapolis Labor Review* covered in detail the failure of Minneapolis city council to address a motion by Tenth Ward Alderman Irving G. Scott to have the city pay for damage done to the state headquarters of the International Workers of the World (IWW) at 14 S. 1st Street on the night of October 20 and to investigate the failure of the police to seek out the perpetrators and bring them to justice. Scott's motion was at the request of the Trades & Labor Assembly. Captain George H. Mallon of the World War Veterans spoke on behalf of Alderman's Scott's motion.

> I am speaking for organized labor. Organized labor has nothing to do with the I.W.W. However, organized labor does stand for law and order and I believe the government and state that I belong to is great enough to enforce it. If mobs are going to be formed in this city, organized labor doesn't know how soon they are going to be used against them. There is no doubt in my mind that the mobs organized have been to further someone's selfish interest. All mobs should be handled without gloves, regardless of the camouflage it uses to conceal its real intent.

Mallon wanted to point out the obvious; i.e., that mobs were out of hand in Minneapolis and throughout the state, as well as the nation. He spoke forcefully from the perspective of a decorated officer of the United States Army to those in attendance, as well as to former soldiers across the state:

> Some of the recent mobs have represented themselves as returned soldiers. No real soldier would disgrace the principles he upheld in the war by joining a mob. It is not the soldier's nature. Look up those men's records and I am of the opinion you will find they were men who never learned self-restraint or discipline and are a disgrace to the men who served their country in her need and when their services were no longer required returned to civil life and took up their duties where they left off. The real soldier's name should be protected by law and not disgraced by a few disorderly characters. I have served my country in three wars and when called upon to serve will gladly do so, but I am sure my Government does not wish me to take the law into my own hands and destroy property ...

Mallon, in order to highlight his point, brought up the matter of the Lundeen kidnapping.

> We had a case yesterday where one of our ex-Congressmen was mobbed by so-called returned soldiers. Such men were hoodlums before they went into the Army and will be hoodlums forever.

Lobbyists, bankers and corporate attorneys were present at the City Council meeting. They "loudly applauded" Chief of Police J. F. Walker, who was unapologetic before the council for allowing the violence. The only sense of law & order came from Alderman Scott. Council even discussed purging itself of Scott. As the discussion became more and more heated, Captain Mallon again took the floor:

> When I came here I didn't know the I.W.W. and organized labor were on trial. The evidence

disclosed here shows there are going to be more mobs unless Council takes action, as certain parties have been located and their addresses carefully noted and mobs are easily formed.

In the end, City Council voted that the city would not reimburse the IWW for the damages and adjourned without making any recommendation as to law enforcement or lack thereof. *Minneapolis Morning Tribune*'s coverage centered on the need to remove Alderman Scott from the council.

Alderman Scott did not back down. He went so far as to suggest that the violence against the IWW was timed to divert attention from the franchise grab of Minneapolis streetcars. *Minneapolis Labor Review* quoted Scott. "They think that if publicity is given the IWW the citizens will not question the $24,000,000 franchise grab so much and not try to discover itching palms in the council of some who are so eager to see the franchise pass." The franchise lobbyists resented Scott's resistance to their monopolies.

In December 1919, the city of Minneapolis put the proposed streetcar franchise up for a vote. The wealthy shareholders of Twin City Rapid Transit Company and of the major newspapers went to considerable effort to denigrate the opposition. That included Captain George Mallon, who came out publicly against the proposed franchise. *Minneapolis Labor Review* (December 5, 1919) printed a statement by the war hero on page 1, column 1.

People Have Too Much Spirit to be Coerced
Inference Cast At Franchise Opponents Resented By
Capt. George Mallon

The citizens of Minneapolis have too much of the real American spirit to be coerced into voting for the franchise by either the press or the Twin City Rapid Transit company . . . the people are not going to reward poor street car service by handing out to the company a one-sided franchise at a grossly inflated valuation which will mean a considerable raise in the rate of carfare . . . It has always and continuously discharged employees who asserted the right of affiliating with organized labor. It has discharged skilled and competent employees who refused to be the serfs of the corporation . . . To be a good, patriotic American citizen we are told we must pay interest in capital that never existed. I believe this attitude will be resented by thousands of service men who served the nation in time of need, but who believe that they were fighting abroad for democracy to return to be exploited by a company, which the *New York World* on Dec. 6, 1917 declared in an editorial did not differ much from alien enemies in the harm it was doing by fomenting labor troubles during the war.

The vote on the Minneapolis streetcar franchise took place on Tuesday, December 9. The *Minneapolis Morning Tribune* headline of the next day read, "Minneapolis Rejects Streetcar Franchise by 6,943 Votes." The *Tribune* noted the very high voter turnout despite the bitterly cold weather. The *Minneapolis Labor Review* headline read, "Magnificent People's Day – Franchise Defeated."

The *Review* went on to report:

The statement of Captain Mallon was of unquestionable influence in the defeat of the franchise and showed the labor and servicemen to be standing shoulder to shoulder. The World War Veterans' organization was untiring in its efforts . . . Tuesday was the day of the people.

Twin City Rapid Transit Company continued to be a strong opponent of labor and an active contributor to the anti-labor Citizens Alliance. The war between the employers and working class was not over.

CHAPTER ELEVEN
NONPARTISAN LEAGUE

Freedom of speech in the United States suffered a major setback at the beginning of 1920. On January 2, the Justice Department launched a series of sudden raids on "radical" organizations in over thirty cities across the United States. Subsequent raids and follow up harassment continued for the next six weeks. Attorney General Mitchell Palmer approved the unconstitutional move and J. Edgar Hoover organized and executed them. Law enforcement did not have search warrants for many of the raids and seizures. They arrested over 3,000 without cause, other than association. Prolonged detention only exacerbated the unconstitutional nature of the raids and arrests. Hoover himself readily acknowledged the brutality involved in the raids. The series of unconstitutional attacks became known to history as the "Palmer Raids." This was the pinnacle of the Red Scare, which had been building since the Russian Revolution during the Great War. The action also forever changed the Justice Department as an agent of power and a reactionary pillar.

In the midst of all the mayhem, the Building Trade Council unanimously re-elected George H. Mallon to be their business agent on January 8. Days later, Mallon was inspecting the construction of a new building in downtown Minneapolis. As he passed a group of workmen, Mallon overhead one of the workers tell another, "There is the officer I picked up in the Argonne." The veteran's name was Frank Mason. He and Mallon shook hands. The next day, Frank Mason introduced Captain Mallon to Bert Hart, the other soldier, also of Minneapolis, who carried him to safety.

Captain Mallon never forgot the courage of the two men, who evacuated him from the front after he was wounded in October 1918. As he was unconscious at the time, he had no idea as to their identity. No one else recalled, as they were all busily engaged assisting the other wounded. Mallon attempted to learn their identity and locate them, but was unsuccessful. It was an important moment for him to be able to thank the two for saving his life.

Whenever Captain Mallon spoke of the war and his decorations, he always gave credit to the men under his command. He insisted that they all received the Medal of Honor presented to him. Mallon often referred to them as 250 of Pershing's Heroes. From time to time, he would interject the brave efforts of Mason and Hart, who carried him from the field when he was wounded on October 1. Captain Mallon was the kind of man who never forgot those with whom he served.

Frank Mason and Bert Hart were both inducted in the military service in Minneapolis in early 1918. Both were assigned to the 88th Division at Camp Dodge. Both were part of the transfer of over two thousand men to the 33rd Division in late March and early April. Private Frank Mason was assigned to Company E, 132nd Regiment, commanded by Captain George Mallon. Bert Hart was assigned to Headquarters Company of the 132nd Regiment. Private Hart saw action at Hamel with Captain Mallon. Private Mason saw first action on September 26, 1918 at Bois des Forges at the onset of the Argonne Offensive. After evacuating Captain Mallon on October 1, Private Mason saw serious action on October 7 and 8 in the crossing of the Meuse River. He later nearly died of influenza. Frank Mason managed to survive, unlike so many millions of others in the fall and winter of 1918-1919.

Captain Mallon and the World War Veterans responded to the Palmer Raids. The front-page headline of the January 23, 1920 issue of *Minneapolis Labor Review* read, "Reign of Terror Condemned by War Veterans. Soldier Organization Protests Intimidation." The headline of an associated article on the same page of the *Review* read, "Service Men Demand Free Speech Press." Three thousand people jammed the auditorium at an event, which the World War Veterans organized. A banner across the top of the stage read: "The Enforcement of the Constitution as It is Written." The great organ played the Star-Spangled Banner to open the event. Carl O. Parsons, State Organizer for the World War Veterans, introduced Captain George Mallon. Mallon was no longer just a warrior, but a respected community leader and a popular public speaker. He addressed the cheering crowd:

> I want to thank you. That means more to me than being one of Pershing's Hundred Heroes. To be a friend of the American people and to know they are my friend is more than the glories of any war can bring to any man. You are my comrades today as much as during the war. We want you to continue to be our comrades until the great fight for democracy has been won . . . It remains to be seen whether you are going to grasp the democracy our comrades died to make possible. There are many breaking hearts in America. I know there are breaking hearts here tonight. But are we going to fight on or stop now and allow the fight for democracy to be lost? I am sure people want democracy, not camouflaged autocracy.
>
> We are what are known as returned soldiers. We hope in America we will be the last, for to have returning soldiers, there must be departing soldiers and always more depart than return. Many brave men must die. Now what we want you to do is see if we can prevent another occurrence of this kind, If American will stand for eternal peace, the rest of the world will follow.
>
> Try to teach love instead of war. If half of the effort and brains were applied to the preventing of war that are to making destructive devices, there would be no wars.

The Next Governor and Lieutenant Governor of the State of Minnesota

Doctor Henrik Shipstead, for governor.

Captain George H. Mallon, for lieutenant governor.

Victory Ticket Named in Minnesota
from The *Nonpartisan Leader*, April 12, 1920

George Mallon, as chairman of the huge event, recognized the opportunity to further another cause dear to his heart; i.e., Irish Home Rule. He invited his friend Reverend William J. Harrington, Assistant Pastor of St. Luke's Catholic Church in St. Paul. Harrington served during the World War as chaplain for the 1515st Field Artillery and was an active member of the American Legion. Mallon introduced the priest by saying:

> Not only did we fight for democracy. We were told to fight for the freedom of the world to govern itself. We believe in freedom for all nations . . . We have a speaker who will talk on behalf of Ireland. Rev. Harrington.

Harrington ". . . paid a splendid tribute to Mallon and declared he deemed it an honor to be with a crowd Mallon was with."

Father Harrington told the crowd:

> Those, who have been supposed to be friends of Sinn Fein, have also been down as Bolsheviks, as the former speaker said you had been. It doesn't detract from them in the least. I know the breed of those who stamped them with that name.

Harrington made clear that "England stole Ireland." He then made an ardent appeal for American aid and support for Ireland and he paid tribute to De Valera, president of the Irish Republic.

Éamon de Valera, a dedicated Irish independence fighter, toured the United States in the latter half of 1919 and through most of 1920. He presented himself as President of the Irish Republic, an entity not formally recognized by the United States. De Valera had been a commander in the 1916 Easter Rising. He escaped from British prison and traveled to the U.S. to build up support from the American people and to secure official recognition of the Irish Republic from the U.S. government. De Valera was controversial for his refusal to fight in the Great War (because of Ireland's war with Great Britain) and for his involvement with Sinn Féin, which many considered to be a terrorist organization. Many Americans felt that made de Valera an enemy. Irish Americans rallied to de Valera's support.

The Irish leader visited the Twin Cities in October 1919. The Twin Cities, particularly the Irish American community, gave De Valera a warm reception on his arrival. More than five thousand jammed into the Armory that evening, cheering wildly and waved American and Irish flags when he arrived. De Valera challenged the characterization of Sinn Féin as anarchists or the equivalent of the Wobblies in the United States. De Valera insisted that there was no Ulster question, only the question of the English in Ireland. At its heart, the matter was the plight of Irish Catholic farmworkers and laborers under the heel of English and Scot Irish landlords and industrial barons. Captain

Mallon led the World War Veterans at the Armory in expressing sympathy for De Valera and Ireland's cause. A banquet at the Radisson Hotel was the final event before De Valera's departure.

Mallon's high profile only heightened the attention of Attorney General Mitchell Palmer and his henchman J. Edgar Hoover in Washington, D.C. Those were dangerous times for anyone who spoke out, but George Mallon did not care.

An internal memorandum reveals how the American Legion perceived Captain Mallon. Harrison Fuller of St. Paul, head of the American Legion in Minnesota, wrote to Frank B. O'Connell, Legion Adjutant in Lincoln, Nebraska on March 15, 1920.

> George H. Mallon. Mallon is a really big man in every sense of the word and is the only force which has held the organization together and kept it from running wild or falling to pieces. He is the man to whom they always "point with pride" when their loyalty is questioned.
>
> Mallon was a Captain of Infantry, winning the Congressional Medal of Honor, Distinguished Service Cross, and several other decorations. He is the Secretary of the Minneapolis Trades and Labor Assembly, I believe, and is, of course, the chief asset of the WWV.
>
> We had hoped through Mallon to bring about the disintegration of the WWV in favor of the Legion, but attacks of other members of the WWV upon the Legion made further negotiations impossible and we gave up the attempt some time ago.

While George Mallon continued to speak and organize, his livelihood and the well-being of his family depended on his work as business agent for the Building Trades Council. That required attention to details, in contrast to the "big picture" look from his national efforts on behalf of the World War Veterans and the Nonpartisan League. In February, Mallon confronted the business superintendent of the Board of Education for Minneapolis. At issue was the awarding of a plumbing contract to a non-union company. The board feared a damage suit if they recalled the contract. Mallon calmly expressed concern about the trouble that might ensue if the Board did not recall the contract.

George Mallon's labor leadership and speaking abilities attracted the attention of Arthur Townley, founder of the Nonpartisan League. Mallon's close friend Thomas Van Lear likely introduced the two. The Nonpartisan League was preparing for the state elections in the fall of 1920. Minnesota, unlike North Dakota, had a considerable manufacturing labor force in addition to a large farm labor force.

In July 1919, William Mahoney and Thomas Van Lear formed the Working People's Nonpartisan Political League (WPNPL) at the Minnesota State Federation of Labor convention in New Ulm. The WPNPL was to progressive labor what the Nonpartisan League was for progressive farmers. The combined farm and labor power faced a far greater chance in the election against the Republican candidates. The two leagues agreed to meet in separate conventions at the same time in St. Paul on March 24 & 25, 1920. The agreement was to confer together in the selection of a slate of candidates.

Brotherhood of Locomotive Firemen & Enginemen's Magazine noted:

> Two years ago, these conventions could not have been held in the state of Minnesota. Or, if by grace of our Governor, Hon. J.A.A. Burnquist, permission had been secured for the holding of the conventions, there would at least have been present the agents of the notorious Minnesota

Safety Commission. The war is over. The Safety Commission is gone. The first rays of the sun of democracy are throwing their light over the horizon of Minnesota. Our soldier boys are back and in the game to see that the democracy that they risked their lives for shall become a reality in the state.

For the 1920 election, Townley presented his plan to the Nonpartisan League to run 700 delegates as candidates for office. A split developed with the Farmer-Labor Party after they created a national party in August. Townley wanted to focus on Minnesota, but was concerned about radicals in the Farmer-Labor Party. He determined that the League candidates would run as either Republicans, if they won the nomination, or as Independents.

In early April 1920, The Nonpartisan League convention nominated Dr. Henrik Shipstead as its candidate for governor and Captain George H. Mallon as its candidate for lieutenant governor. The lieutenant governor was an important state official because he presided over the state senate and, perhaps more importantly, appointed the committees. At risk was whether labor or the steel trust would appoint the committees.

Mallon was well-liked and known as a good public speaker. He also was a fighter and the League needed fighters. George Mallon was just as valuable an asset for the Nonpartisan League, as he was for the World War Veterans, in the sense that he was a war hero. Red-baiting was far less effective when the accusers had not even gone to war and were charging a Medal of Honor recipient with disloyalty.

Mallon, as candidate for lieutenant governor, did show his character and the courage of conviction of his beliefs, although perhaps also a bit of political naiveté. The League debated the merits of the British government's position as it related to Ireland. The British favored a home rule measure, which kept Ireland as part of the United Kingdom. Mallon and others favored self-government for Ireland; i.e., an entirely independent Ireland. George Mallon never forgot his Irish roots. His papers include a copy of an old yellowed copy of an Irish national newspaper with articles such as "Sinn Fein Man Escapes J. Bull's Jail" and "De Valera Scores British Tyranny in Ireland".

The headline of the April 24, 1920 issue of Minneapolis-based newspaper *The Irish Standard* read:

WHICH STAND WILL AMERICA TAKE?
Which Will It Recognize?
The Alien Government of Might or the Native Government of Right?
The One Has No Sanction But Brute Force,
The Other Has The Supreme Democratic and Moral Consent of the Consent of the Governed

The Irish Standard published an op-ed on June 19, 1920 titled, "No Ulster Question, Says Griffith." Arthur Griffith, Acting President of the Irish Republican Government insisted that the English were propagating a myth that the six counties in the north of Ireland were opposed to independence. Griffith related incidents of coercion and arrest in Mallon's ancestral County Tyrone of local leaders, as well as confiscation of their registers. Despite a strong majority of Irish Catholics in County Tyrone desiring independence, the English were not allowing democracy to find its true course.

Minnesota Scandinavians had little to no interest in the matter of Home Rule for Ireland. However, George Mallon retained his deep connection through his father and grandfather to Ireland. He was not alone in Minnesota. St. Paul, in particular, was heavily Irish American. But the Twin Cities, where George Mallon lived and raised his boys, was far different from rural Minnesota. Nonpartisan Leaguer W.C. Hedenberg of Minneapolis opposed taking a stand for Irish self-rule. In his opinion, the League's position in favor of self-government for Ireland could hurt the League's efforts to attract votes. He specifically mentioned some veterans in Norman, Kittson and Polk Counties in the western and northwestern parts of the state opposed self-government for Ireland. That was a point worthy of discussion in a political campaign. George Mallon rebutted:

> Are the World War Veterans catering to intelligent people or to a few ignorant Swedes? Everyone who knows history and knows anything about the war for democracy should know that Ireland is entitled to the same right of self-determination as the other nationalities of Europe.

Mallon was on the right side of history. However, whether or not anyone knew that, his first sentence infuriated Minnesota Swedes.

The opposition did their best to highlight Mallon's comment and inflame voters. Newspapers across the state published editorials and letters expressing outrage. Minnesota Issues, a well-funded conservative newspaper, made a point of re-publishing the most outrageous. Among those was from The *Owatonna Chronicle*:

> The "weak sister" among world war veterans' organizations, the "World War Veterans," held its first annual convention in Minneapolis last week . . . and attempted to impress the Northwest with the idea that war veterans want to tip the country upside down, apply the torch and then aid the Communists, Nonpartisan League, Socialists and others of the breed to loot it . . . Holding a convention in Minnesota and listening to a hot tirade against "ignorant Swedes" by the Nonpartisan League's candidate for governor seems hardly consistent with a minute degree of common sense . . .

Minnesota Issues itself commented on "Mallon's vicious attack" and referred to him as "the typical agitator." There is little doubt that Mallon cost himself the Minnesota Swede vote. Just as ethnicity mattered to Irish Americans, it also mattered to Swedish Americans. Neither George Mallon nor the Nonpartisan League backed down.

The League did its best to run large advertisement reads heralding Mallon's heroism in the service. The ad read in big bold letters: "Elect Captain George H. Mallon Lieutenant Governor." Smaller headers read: "Vote Here as Mallon Fought There – For Liberty and Democracy" and "Save Minnesota from the Grip of Organized Greed." Central to the ad was a reproduction of the sketch from the Liberty Loan ad, showing Mallon throwing a punch into the face of a German officer under a banner, which read "America's Immortals." Underneath read: "The Test of Our Americanism", a rebuttal to the corporate interests, who argued that Mallon's support of decent pay and working conditions for the working class was "radical" and "un-American" and even "Red". To the right of the sketch was a large formal portrait of George Mallon in a suit and tie. Newspapers controlled by big business refused to run the advertisement.

The *Nonpartisan Leader* of April 12, 1920 quoted Mallon on another sensitive matter under attack:

> While I served my country abroad during the war, that does not entitle me to any special consideration at your hands. Many men and women, laboring in the cities and on the farms, served their country just as truly as we did in the army. In one matter especially, I do not want to pose under false colors. Although I served my country in two wars, I am not in favor of compulsory military training. It is a form of Prussianism and I am opposed to it . . . to my mind compulsory military training is no part of democracy . . .".

Emotions were running high across the state. Opponents of the Nonpartisan League made every effort to employ Red-baiting to prevent George Mallon and the others from becoming candidates. Billboards across Minnesota and Republican-owned newspapers carried a simple message: "Stop Socialism." J.A.O. Preus, Republican candidate for governor, focused through the campaign with conditions in Russia, trying to somehow tie that to the Nonpartisan League. Louis Collins, George Mallon's opponent, as well as his veteran colleague and friend, told a Republican crowd:

> This fight is one that involves the church, the home and the state. The war isn't over when the voters are confronted at the polls with candidates picked by men who sit at the same table with convicted disloyalists and choose as their associates and advisors men high in the circles of radical Socialism.

Minneapolis Labor Review, in turn, was furious when supporters of Louis Collins circulated a pamphlet citing a three-year-old article from the *Review* about Collins being not such a bad guy. The headline of the rebuttal read in bold black letters, "Labor Movement is Not Indorsing Louis L. Collins." The editor of the *Review* made his point in no uncertain terms:

> In this campaign, we are dealing with men as they are today, not as we thought they were years ago. Captain Mallon came out of the war the most lavishly decorated soldier in Minnesota . . . unlike Collins, he is continuing at home the fight against capitalistic militarism, which he and millions of others believed they were participating in when serving abroad. Steel trust politicians stole the victory from the men who served in France and, while Mallon fights the steel trust and the enemies

of the people at home with all the courage and vigor which he displayed in France, Collins consents to become a candidate for lieutenant governor on the steel ticket headed by Jake Preus.

On May 23, Mallon told a crowd of several thousand assembled in New Ulm that he wished more mayors were like New Ulm's. He related instances of hostile receptions in numerous towns across the state. At the same time, he assured the crowd that he always held his meetings. He deplored the situation in an alleged free country in which mobs prevented a man from speaking his mind. He commented on bringing an escort of World War Veterans, if necessary. *New Ulm Review* (May 26, 1920) dedicated several columns to the event and the contents of the speeches.

> The entire credit of his great military success Capt. Mallon gave to the privates under his command, stating that the officers would be absolutely helpless without the true devotion of their men. He especially lauded the lads who, ignoring their own danger, carried him, wounded, from the field of battle.

The service of enlisted men in the war was a common theme of Captain Mallon's talks, which inevitably led to:

> Our Doughboys had been told that they were being sent to Europe to fight for democracy ... the very thing they were sent overseas to help eradicate had sprung up behind them, even more insidiously than it had ever flourished in Europe ...

Mallon rarely took that a step further, but he did so in New Ulm:

> The men came back full of enthusiasm after having been victorious in Europe, but this enthusiasm waned when they became aware of an attempt to organize them into an un-American association, the hidden purpose of which was to fight the organized farmers and labor. The ex-servicemen are determined to place the government back into the hands of the common people . . .

He did stop short of naming the American Legion, but everyone knew to whom he was referring. As Mallon had become fond of telling crowds, he said he would rather be sent to the rock pile than have to occupy the pile of marble in St. Paul without the other candidates in the League-Labor ticket. He urged voters to understand that excess profits tax legislation would eliminate war profiteering and certainly would lessen the likelihood of war.

The heavily German-American community of New Ulm suffered during the war. Many spoke German, read locally published German language newspapers, attended school in German, went to church for Mass conducted in German and much more. More than 10,000 people attended an anti-draft protest rally in New Ulm not long after the U.S. entered the war. There were speakers in 1917 speaking, as Mallon did in 1920, about the profiteers pushing America into the war. The result was the arrest of some leaders and the expulsion from employment of others. Mallon not only supported the profiteering argument. He endeared himself to the crowd by stating unequivocally that he investigated and he felt that the "blood-curdling stories of alleged atrocities" were propaganda.

Mallon was a man with a sense of humor. He closed his engaging address by facetiously telling everyone:

> . . . to subscribe for all of the big Twin Cities dailies, if their funds permit; to read them carefully each

day and, when election day rolls around to vote just the opposite from what these papers advocated.

Captain Mallon traveled all across the state and continued to hammer away at the opposition and appeal to the working class, both labor and farmers. At Glencoe on May 28, he offered to a crowd:

> I would rather fail on the League and labor ticket than be elected governor on any other . . . Your chairman has seen fit to refer to me as having won these decorations. I want to assure you that I do not and never have taken the credit of one of those decorations. They were won by the men who served under me — American soldiers — enlisted men. No officer in this United States army could accomplish what we did unless he had had the right spirit among the men; unless he had men, men that were soldiers, men that came from the common people — labor boys and farmer boys. Boys of all walks of life, men of all nationalities, welded together in this country, went to Europe to fight for democracy. I really thought when I went over to Europe that I was told the truth when I went over and was told that I was fighting for democracy. Upon my return, I discovered that we had not won for democracy — that we had built up in this country a system of autocracy that was far more dangerous to this country than foreign militarism ever dared to be.
>
> Now on my return from service, I tried to get rid of all the military feeling that I had. I don't think that when you look at me today that I look like much of a soldier and I don't want to feel that way. I want to feel that I am back again as a workingman.

> I was born and reared on a farm. I came to the cities and learned my trade and have worked at it continuously ever since. When I returned, I took up my work where I left off.

Mallon was setting up the crowd for a strong upper cut to the jaw. Again, he returned to his rock pile remark:

> ... if there is anybody here that is going to vote for me and not vote for the rest of the ticket, I am going to ask you not to do it. I would sooner be sent to the rock pile than to that marble pile alone. If I should get in there alone among that bunch of highbinders it would not be long till they would have me in Stillwater some way.

> Now you know when I came back to this country, there were quite a few new sayings... "One hundred percent American" ... it was quite a while finding out what "100 percent Americanism" was ... a steel corporation's earnings ... in one year during the war ... they made 100 percent profit. Then I knew what 100 percent Americanism meant. If this system is allowed to go on much longer a man will have to be 1,000 per cent American to be a good citizen.

Minneapolis Morning Tribune announced in its May 30, 1921 issue that 15,000 would march on Memorial Day, making it the largest parade in Minneapolis history. Captain George Mallon and Andrew G. Cooper led the sixth division, the World War Veterans. *Minneapolis Labor Review* offered commentary a few days later on page 1 of its June 4 issue under the headline "World War Veterans in Memorial Parade." The Review noticed a more important first:

> The veterans of '61 and the World War Veterans' organization were the recipients of the greatest applause along the line of march Memorial Day . . . Led by Captain George H. Mallon, who refused to ride either horseback or in an automobile, but insisted on walking with the rank and file of the organization, the World War Veterans presented a most impressive appearance

The Review noticed "lack of enthusiasm and applause."

> It seemed as if the crowds were stunned with the display of artillery manned by the state militia which led the parade. After millions had sacrificed their lives to end all wars, the people were not at all sympathetic to having displayed in a Memorial Day procession more machinery of destruction than had ever been exhibited on a like occasion.

An editorial in *Minneapolis Labor Review* on June 11, 1920 pointed out, "... in the case of the lieutenant governorship, it is practically a man to man fight between Mallon and Collins." The Review reminded all that Mallon always commented:

> Every time I look at these medals, I realize that they belong to the men who fought with me, just as much or more than they do to me, and I will never forget that they are baptized with the blood of American soldiers.

The Review commented, "A man who doesn't forget the men who fought alongside of him can be depended upon not to forget the people who elected him." This was and is the essence of democracy.

Nonetheless, Mallon's opponent was a formidable one. Louis L. Collins was a newspaperman, a war veteran and a man as well liked as and perhaps even better known than Mallon. Collins' father, an early Minnesota settler, became a prominent attorney and, later, a justice on the Minnesota Supreme Court. Collins also had considerable support from comrades with whom he served during the war as part of the Minnesota National Guard's 151st Field Artillery.

On June 14, as the Republican primary neared, *The Nonpartisan Leader*, the League's newspaper, optimistically declared: "Minnesota Victory Campaign near End. Primary Election June 21 to Sound Knell of Corporation Control of State Government." There always is, after all, some inherent mystery in election outcomes. The full-page article included a large reproduction of the Victory Loan poster featuring Captain Mallon throwing his famous punch into the jaw of the German officer.

Two days later, The *New Ulm Review* reported on George Mallon's speech at nearby Lamberton on June 11. This was after the mayor declared that he would not allow Mallon to speak in his town.

> The people of Lamberton had an opportunity last Friday to hear some first-hand information along progressive lines, not only politically, but also on true Americanism, when Captain George H. Mallon, candidate for the vice governorship of the state, addressed a large crowd. It was one of the best political meetings ever held in Lamberton and Capt. Mallon was given a hearty welcome by the people of Lamberton and the surrounding area. The meeting was held in the open air.

In the midst of the Minnesota primary elections of June 1920, an event took place, which shocked the state and the nation. A Duluth lynch mob hung three African American carnival workers. This heinous event foreshadowed the rise of the Klan in Minnesota in the 1920s. *Minneapolis Labor Review* offered an interesting perspective in an editorial titled, "Minnesota's Bid For Shame."

> So long as colored men are denied the right to fair and impartial trial, no one can boast very loudly of the efficacy of law and order. The steel trust has for many years carried on a policy of lynch law against the workers of the iron range. In this way, they have sown the seeds of outrages like the one which occurred Tuesday in Duluth.

By June 1920, as the Republican primary loomed, Mallon had his gloves off and was throwing some bare-knuckled punches into the face of the opposition. He had little stomach for tolerating cowards who took part in mobs. The Willmar *Tribune* of June 9, featured a headline, which read, "Yellow Paint Boys Flayed by Mallon. "Scabs" Might Have Painted No Man's Land, Candidate Says." Mallon told the crowd of supporters:

> We want to know who the scab painters were who stayed at home to smear yellow on the homes of innocent citizens while the American boys were painting No Man's Land red with blood. If the men, who under cover of darkness, painted men's homes yellow, wanted to do some right, they could have found plenty of it in No Man's Land.

> I know what war is. I have been through three of them — the Spanish-American War, the Philippines War and the World War — and I am willing to put on the uniform again if it will help finish the job of smashing militarism.

> I am in favor of putting laws on the statute books that will take all the excess profits out of war.

Then there will be no more wars. The common people of this country have no grudge against the common people of any other country.

There are several things which the returned service people want to know. They want to know how it is that some men remained home and made millions while others were getting shot to pieces at $30 a month. We want to know about the profiteers who waved the flag with one hand and with the other hand took 50 cents of your Liberty bonds out of the government's pocket.

The Nonpartisan League slate spoke to a large, cheering crowd at the Minneapolis Auditorium on Monday evening, June 14, 1920. The still air was oppressively hot, but there were no seats available for three hours of speeches. Tom Van Lear facilitated the event. *Minneapolis Labor Review* (June 18) reported, "The greatest ovation of the evening was accorded George H. Mallon." Mallon was at his best, highlighting patriotism and injustice. He opened with a bit of humor, a subtle nod to the Irish struggle for independence:

> Your chairman saw fit to compare me to a bulldog. If he had called me an English bulldog, I would have been offended.

Mallon moved quickly to the matter at hand:

> I don't think the workers and laborers need any education in patriotism.
>
> I wouldn't mind being called un-American by the Citizens Alliance. I'd rather they call me anything than to call me their friend.

Then Mallon hammered his main point. His exploits as a war hero earned him the right to speak to the origins of war:

> The profiteers have made their millions while you and your comrades were sacrificing their lives. For every two men left in France this country made a millionaire. I could never feel that I was keeping the trust of those men unless I stood for what they died for. That flag. That flag means the common people. The Kaiser of Germany was the same as the kaisers of industry here. While we were knocking over the Kaiser, they were building them up here . . . before we go again we will make sure what they mean by democracy. We will make them cut the profits out of war and when the profits are cut out, wars will be cut out.

It was this direct attack on the system of Big Business and its control of the government that resulted in the many attacks on Captain Mallon. To this end, Mallon said:

> They have a Sound Government League. I am for sound government, but there are some sounds about their kind of sound government that I don't like. I don't like the sound of their jail doors clanging behind the class prisoners, nor the rattle of their machine guns shooting down workers.

Captain Mallon addressed the matter of his friendship with Republican opponent Louis Collins:

> I have never yet knocked my opponent. He and I have the same opinion of each other. We are personal friends, but each thinks the other is in bad

company.

There were those, who did their best to challenge the Red-baiting. The Duluth newspaper *Labor World* published a letter from Bob Morgan of the Plumber's Union on June 19, the day before the primary.

> I do not care what others may say, I know that Cap Mallon is not red. He is just straight red, white and blue and he showed it in the big fight across the pond . . .They say he is all right, but they don't like his company. Well, they weren't so particular when they needed a good soldier and I guess we are as badly in need of good statesmen today
> Though one of Pershing's hundred heroes, and the most decorated man in Minnesota, he is the same old Mallon that he was when we cut pipe together. "Had Mallon been endorsed by the old guard the newspapers would be filled with stories of his wonderful war record."

Of course, the opposition neither subscribed to nor read *Labor World*. And as *Minneapolis Labor Review* pointed out, "Had he been one of the exploiting class, he would have received pages of comment in the daily screeds." Instead, the major St. Paul and Minneapolis newspapers ignored the distinguished candidate.

The Nonpartisan League candidates failed across the board to win any nominations in the Minnesota Republican Primary held on June 21, 1920. Louis Collins defeated George Mallon for the nomination as lieutenant governor by 17,000 votes. Jake Preus, Republican candidate for governor, declared, "The campaign just ended, which was led by Corporal Collins and myself, was to prevent a coterie of Socialists from getting control of the Republican party organization." In response, the Nonpartisan League announced that its slate of candidates would run on an independent ticket against the Republican and Democratic candidates in the November general election.

The *Nonpartisan Leader* of July 5, 1920 pointed out the obvious:

> The workers and farmers were unable to match dollar against dollar with the steel trust. But even had they been able to do so, it would have been of no use. Personal friends of Captain George H. Mallon contributed to a fund to place a half-page advertisement telling of his war record and reproducing the Liberty loan poster... The *St. Paul Pioneer Press* and *Dispatch*, the *St. Paul Daily News*, the *Minneapolis Daily News* and the *Minneapolis Tribune* refused point blank to accept this advertising matter, showing the venom of these sheets.
>
> But while the bitter attacks upon the League and labor candidates apparently accomplished some result in the cities and towns, they fell flat in the country. In spite of rains and muddy roads, the farmers everywhere swarmed to the polls and registered their votes for democracy and against the steel trust.

The Nonpartisan League, defeated in the Republican primary, proceeded to its next move, that of filing to run as an independent third party in the general election in November. Once again, George Mallon would face Louis Collins. The *Brotherhood of Locomotive Firemen and Enginemen's Magazine* noted:

> Great Gains Mark Political Progress of Producers' Movement at Recent Primaries. The reactionary press in Minnesota has congratulated itself on the alleged defeat of the Farmer-Labor forces in the primary election in Minnesota on June 30, but as

usual the reports were colored and misleading. It is true that we lost the state ticket to the old guard candidates, but the margins were so close in some places that the old guard had little to crow over when all the votes were counted. The kept press had much to say about the "defeat," but touched lightly on the victory of the Nonpartisan League . . . We are still in the race . . . the reactionaries will be defeated if the progressive votes in the state are as much in evidence as they were in the primary election.

The League closed the margins in nearly every race. The League also placed candidates on the Republican card for the Supreme Court, two District Courts and state legislatures in 78 of 86 counties. The *Brotherhood of Locomotive Firemen and Enginemen's Magazine* added:

> The Farmer-Labor forces were badly handicapped by the lack of newspaper support and in St. Paul and Minneapolis newspapers refused to publish advertisements containing the war record of Capt. Mallon.

The *Labor World* in Duluth announced the filing of the Nonpartisan League as independents for the November election. "Their entry in the campaign has the old gang guessing as it never guessed before." The paper noted how narrow the margin of defeat had been for the three candidates and stated, "Captain Mallon, a novice in the political game, showed up remarkably well in the primaries." The article quoted Dr. Shipstead, the gubernatorial candidate:

> The Steel Trust beat us in the primary skirmish by the use of large amounts of poison gas. Their America Committee, Citizens Alliance and Sound

Government Association carried on a scurrilous propaganda of slander the likes of which I doubt has ever been equaled in any campaign at any time in the United States. The air of Minnesota was polluted with their foul breath . . . We expect to make this fight so hot that the iron heel of the Steel Trust will be melted and made into steel pens with which the people of Minnesota can write a new Declaration of Independence.

The Republican slate received some strong support when the Republican candidate for president, U.S. Senator Warren G. Harding of Ohio, made an appearance at the Minnesota State Fair on September 8. On the stage with him was Governor Burnquist, Republican candidate for governor J.A.O. Preus and other Republican candidates and dignitaries. Over 82,000 were at the fair that day with 30,000 in the grandstand. Harding's address also was broadcast over the radio across the state – a first for a presidential campaign. Harding addressed his remarks to the farmers of the Upper Midwest. He argued for greater farmer representation in government, the right to form co-operatives, research to study the relationship between crop pricing and production costs, promised to put an end to price-fixing, outlaw pricing speculation and administer the farm loan act. Harding's proposed program, as was intended, severely undercut the Nonpartisan League's edge over Minnesota Republicans.

George Mallon took time off from the campaign trail to attend the American Legion's 2nd Annual National Convention at Cleveland, Ohio on September 27-29, 1920. Camaraderie overcame political differences and Captain George H. Mallon while not a member, was among those present and honored. A photo of Mallon and other Medal of Honor recipients at the Legion convention appeared in newspapers across the nation. Among many good friends photographed with Captain Mallon was his First Sergeant Sydney Gumpertz. Gumpertz settled in

New York after the war and joined the S. Rankin Drew Legion Post in New York City's Theater District.

The program for the Legion convention included an article titled, "What Makes Men Red" by John W. Love. One insightful comment was:

> A very careful study was made of the relation between brains and radicalism. Figures clearly indicated that radical beliefs are more common among classes with higher mentality and practically do not exist among those who cannot read or write . . . cast some doubt upon our generally accepted ideas that education by itself is a remedy for radicalism.

In fairness, Love included some interesting responses from "radicals" as to how they interpreted "100% Americanism", which was an important tagline for the Legion:

> Americanism means the immigrant must understand the American, but not that the American should understand the immigrant . . . Americans expect Americanism from us, but they don't believe in it themselves . . .
> Americanism is opposition to free speech.

The general election, as in the primary, was a war of words. *Farmers Independent* published an editorial on October 7, 1920:

> George Mallon is a fighter, not merely in war, but a fighter for the right to peace. He is one of those plain, honest, "common" men whose heads are never turned by honors. He has not sought to trade upon his reputation as a soldier – and he is a veteran of two wars – but since his release

from the army he has been working for the cause in which he believes as a trades union organizer. George Mallon as a soldier was one of the type of which Americans can be most proud, a man who abhors war and considers it to be the first duty of a good citizen to make wars impossible, but who, when he believes the cause is just and that a man should fight, is ranked among the bravest, the coolest-headed and the most effective of warriors. He is one of the type of men which forms the very backbone of a democratic form of government.

Farmers Independent lamented the "Kept Press", noting that not one of the other papers in the county carried more than a couple of lines at a Nonpartisan League gathering, even though "the largest number of people ever gathered in the county were at that meeting."

In the final week of the campaign, *New Ulm Review* of November 3 offered an unusual perspective in advocating for Mallon in the lieutenant governor race despite the attack of another newspaper. The editorial piece says much about the subtle aspects of the race, beyond the party delineations:

> Our local contemporary attempted to belittle the military record of Captain George Mallon, Farmer-Labor's Independent candidate for lieutenant governor by admonishing its readers to vote for "Louie" Collins, the "little corporal," thus "giving the man in the back lines a chance for once." Judging from the way "Louie's" military achievements were lauded during the pre-election campaign, the "little corporal" was not in the "back line" at all, but exhibited great courage and bravery. On the other hand, it can be states that, regardless of all attempts to belittle and malign Captain Mallon's military record, he rose from the ranks and his

promotion was well-deserved, as witness the fact that he us one of General Pershing's 100 heroes of the war. This enviable distinction was gained through sheer bravery in action, not through being a "swivel-chair" officer.

The Minnesota statewide election was held on November 2, 1920. The deflationary recession of 1920-1921 had been underway for ten months. That gave the challenging Republican Party the edge. Warren G. Harding won a landslide victory for the office of president and that carried over into the states. Republican candidate J.A.O. Preus with 53% of the vote defeated Nonpartisan League candidate Dr. Henrik Shipstead with 36%. Republican candidate Louis Collins with 57% of the vote defeated Nonpartisan League candidate George Mallon, who received 30%. One newspaper summarized the election for Mallon:

> Mallon . . . accepted the nomination for Lieutenant Governor . . . that he knew meant practically sure defeat . . . and rejected the offer to run for governor on the Republican ticket that meant unquestionably election, honor and new fame.

George Mallon was a man of principle, who was less concerned with personal success at the cost of his beliefs. He wanted to see the government be of the people, by the people and for the people, as the Founding Fathers intended.

The pro-Nonpartisan League *Kansas Leader* of Salina attempted to put the 1920 election results into proper perspective; i.e., that the League gained seats in the Minnesota Legislature.

Among fragmentary election reports you noticed probably, statements that the Nonpartisan League had suffered crushing defeat everywhere ... The truth is that the Nonpartisan League made very large gains in strength, consolidated most of its previous positions and showed remarkable solidarity ... in spite of the Republican tidal wave ... In Minnesota, the league considerably increased its representation in the Legislature. Its candidate for Governor, Shipstead, received 275,000 votes, more than doubling the vote given him in the June primary. His successful opponent, Preus, befriended by the Steel Trust, was elected by considerably less than the plurality given by the same trust's successful candidate for governor two years ago. In Wisconsin, where the league had virtually no strength in 1918, it elected the governor, Blaine, by more than 100,000 plurality and a considerable legislative delegation ... In Montana, Colorado, Nebraska, Washington, South Dakota and Idaho, the actual vote in virtually every instance shows large league gains. The fact is that this expanding political organization, altho only a factor in a few states, polled a total of more than 1 million votes – three times its 1918 strength – in the very hour of the Harding landslide.

The Nonpartisan League gained enough ground that Republicans attempted to repeal the Minnesota law, which provided for rank-and-file voters to choose their candidates in the primaries. The intention was to shift the selection of candidates to a backroom deal by the powers so that they would not have to worry about a similar problem in the 1922 primary and election. They were unsuccessful.

In the aftermath of his unsuccessful run for state office through the Nonpartisan League, George Mallon became involved in organizing the history of the state's participation in the World War. The state formed the Minnesota War Records Commission to collect and preserve information as to Minnesota's role in the war. In December 1920, the Minnesota War Records Commission formed a Hennepin County War Records Commission. The executive committee was comprised of George E. Leach, George H. Mallon and Miss Gratia Countryman, librarian of Minneapolis Public Library.

The National Executive Committee of the World War Veterans in January 1921 circulated an overview of 1920 and an appeal to trade union locals across the country to become more involved.

> The men of the World War Veterans, 90 percent of whom went through fire and blood for America and humanity on the pledge of the rulers of America that after the war America would be made right, propose to hold those rulers to that pledge. For them to declare industrial war on the men who fought through the bloody hell of France is to slap us in the face, for we are the men of labor and in a fight, we do not know how to turn the other cheek; so the World War Vets have made labor's fight their own and we want labor to make our fight their own.

The letter included several examples of what could be accomplished through cooperation between the World War Veterans and organized labor. Minneapolis was one cited. While Captain Mallon's name was not mentioned, his impact was woven throughout the story of Minneapolis in 1920.

> Take another instance. Labor of Minneapolis was facing a crisis. The political powers of the city and state, combined with all the other forces and powers of industrial autocrats, were delivering

telling broadsides against Labor. They were refused permission to hold parades, meetings were broken up by the Blue Coats, leaders were threatened with jail and were thrown in jail for unconcealed contempt of court process. The World War Veterans sprung into the breach, for the Vets enjoy nothing so much as a clean, open scrap for the rights of mankind. We shot a bit of Nervine into the Labor men and got them marshalled by the thousands, but were refused permission to hold a parade at the same time that the other gang was staging one. We told the City to go to Hell, and if they tried to stop our parade we would kick their bullies into the mud (one of our posts there number some 5,000 hard-boiled ex-doughboys). We held the parade, and the rough stuff vanished. Minneapolis is a fairly decent American town now.

While the tone of the National Executive Committee was optimistic, it remained to be seen what would happen to the beleaguered World War Veterans in 1921.

CHAPTER TWELVE
RETURN TO KANSAS

George Mallon was out of the political ring for only a short time. The Nonpartisan League recognized Mallon's courage and determination to fight, as well as the qualities that made him a leader of men. The League named Mallon to be State Manager of the Nonpartisan League in Kansas at the beginning of 1921. Mallon believed fiercely in what the Nonpartisan League was trying to accomplish.

> We are here to get the farmers to organize. We are here to help them cast their ballots in their own favor. We intend to stay here as long as the farmers want us. We will stay within the law.

As always, Mallon was determined to see real democracy implemented without the undue influence of the privileged.

The Nonpartisan League knew what they were doing in bringing in Mallon. Kansas remained a battleground and was not for the faint-hearted. J. O. Stevic, the previous state manager resigned six months earlier. The *Topeka State Journal* published his letter to Governor Henry J. Allen on page 3 of its June 10, 1920 issue.

> On the first day of June 1920 in Ellinwood, Barton County, Kansas, there was a picnic held by the farmers of that and adjoining counties for the purpose of becoming acquainted and listening to an address by a speaker for the National Nonpartisan League . . . 300 men, representing themselves to be ex-service men in the United States Army, came into the grounds and seized Walter Thomas Mills . . .

Walter Thomas Mills, who stood four-feet-six-inches tall, was a renowned orator in his mid-sixties. He joined the Nonpartisan League because he believed in the mission.

> A scuffle ensued between the farmers and the alleged ex-service men and Mr. Mills was rescued from the mob . . . the sheriff of Barton County arrived on the scene. The sheriff was requested to give the aged speaker and his party protection from the mobsters, but he refused to do so or to swear in deputies to assist him in maintaining order. On the other hand, he ordered the speaker, Mr. Mills, and his friends out of the park and forced them to leave their friends and take to the highway, where the mob surrounded their car, took charge of it and drove with them to Great Bend, where they marched thru the streets, beaten and robbed, herded in the stock pen, egged and made to suffer other indignities with the knowledge of the peace officers of Barton County.

Mills' kidnapping and assault was not the first such crime against Nonpartisan League speakers in Kansas. The *State Journal* reported in January 1920 of a case in Stafford in which, "The trial of the alleged kidnappers of Nonpartisan league organizers really developed into trial, instead, of the Nonpartisan League itself." Papers like the Hays Free Press referred to, "this vicious trinity – the Nonpartisan League, the IWW and the Bolshevikis."

Townley sent Mallon to Kansas in preparation for the League's big push into the state. Farmers and laborers in Saline County – sixty miles southwest of Mallon's hometown of Ogden – invited Townley to "push a drive for membership." They had the support of Sheriff Ernest Swanson, who issued a statement that he would not tolerate mob violence. A headline on page one of the January 3, 1921 issue of *Topeka State Journal* read, "Legion

Men of Ten Counties Take Part in War." The article reported that men were heading to Salina from Great Bend, Abilene, McPherson, Concordia and Junction City. The article also listed the names of sixty-three Nonpartisan League men in the area. It seemed that the paper, so as to not appear to be targeting the men, cautioned its readers to be aware that they were "skilled in the throwing of the Bolsheviki propaganda."

The *State Journal* of January 5 printed the text of a telegram, which John N. Floyd, chairman of the Americanization committee of the Kansas State Legion, sent to the Legion Post in Salina Kansas:

> Every member of every post in the state is watching with approval your actions and the stand you have taken on Townleyism and the Nonpartisan League. I congratulate your post on its fight and offer you the support of the Americanization committee if we can help you.

Kansas was Mallon's home state. Even as Legion men headed to Salina, the *Junction City Daily Union* of January 8, 1921 reported that Mallon "... is well known in Junction City. He is a member of the Mallon family of Ogden." There is no record if the Junction City boys stopped by to say hello to George Mallon in Salina.

News of Mallon's involvement in Kansas spread throughout the Midwest. The *Kansas Leader*, published in Salina, headlined, "Captain Mallon is Here!" in its January 6, 1921 issue and emphasized, "His connection with the League as an active worker is refutation of the charges made here by the American Legion that the League is disloyal." *The Bismarck Tribune* of January 7 reported. "Nonpartisan League forces in Salina have been strengthened by the arrival of George H. Mallon, former army captain of Minnesota, who came especially to assist A.C. Townley ...".

The editor of the *Kansas Leader* wrote in an editorial:

... it was some little time before we identified him as the George Mallon whom Pershing selected among the American "Immortals" for exceptional bravery and heroism in France.

He taunted Nonpartisan League opponents and honored Mallon as the warrior that he most certainly was by writing: "Personally, after his record in France, we are not going to be in any mob that tries to stop him."

Nor was coverage restricted to the Midwest. The attack on free speech was too important to not cover the events. *New York Tribune* of January 11th carried an article headlined, "Veterans Open Fight on Non-Partisans, Former Soldiers Declare League is Un-American." The *Tribune* read, "Nearly 500 former service men, representing cities in every section in Kansas, met here this afternoon and effected an organization to be known as the American Defense League of Kansas" and identified the leader as O. A. Kitterman, commander of the Saline County American Legion Post. The *Tribune* also cited "Captain George H. Mallon" and his war decorations, as well as Mallon's insistence that the organization was not un-American. The *Washington Post* also carried the story.

The January 13, 1921 headline of pro-Nonpartisan League *Kansas Leader* read: "Crowds Greet Townley. Convention Hall Capacity Inadequate." The event was the first speech by Townley in Kansas. Farmers in Kansas had been writing Townley for some time to convince him to travel to their state to help them organize a Nonpartisan League, such as he did so successfully in North Dakota. The article announced, "Four League Speakers Answer Charges of American Legion and Chambers of Commerce, Tell Why the League is Being Opposed and Audiences Leave with Favorable Impression." Among the four speakers was Captain George H. Mallon.

At issue was whether the Nonpartisan League was about progressivism or radicalism. The farmers sought economic justice and the wealthy intended to maintain as much of their wealth as possible. The opposition, as always, did their best to taint progressives as radicals, Reds and Bolsheviks. The *Kansas Leader* of Salina offered a defense to the charges of Bolshevism in its January 27, 1921 issue:

> The National Nonpartisan League owes its existence to an economic theory of society, rather to anything political, although it employs the political machinery of a state in order to carry out its economic program. Like almost all economic theories, its germ was created as a result of unjust economic conditions which made the life of its members unbearable, if not impossible, to endure. It is in no sense partisan, either in name or in practice, since it supports both Democrats and Republicans in all elections. All it asks of those whom it nominates and supports with its votes is to carry out its economic program; whether these call themselves Democrats or Republicans does not matter . . . It is its nonpolitical, and more particularly its nonpartisan, character that gives to it the strength and chance for success which all politicians so much fear. Doubtless, much of the opposition and much of the criticism of the organization are based on this fact.

The national executive committee of the American Legion met at the Raleigh Hotel in Washington, D. C. on February 7 – 9. Among the matters on the table for discussion was a request from, among others, O. A. Kitterman of the Salina, Kansas Post, that the Legion go on record as opposing the Nonpartisan League. Kitterman also requested support from the national office in his bitter fight against the League. The *Hays Free Press* of February 17 reported extensively on the committee's response.

The committee then voted unanimously to uphold F. W. Galbraith, national commander, in his stand in advising state commissioners of Kansas, Nebraska and Oklahoma that the Legion should take no active part against the League as an organization . . . contended that such action would be considered by many farmers and other members of the League, who themselves were loyal, as indicating opposition of the Legion to the economic and political policies of the League.

Frederick W. Galbraith was a colonel in the world war, commanding the 147th Regiment/37th Division in France. He received the Distinguished Service Cross for his heroism on September 29, 1918 during the Meuse-Argonne. An Ohio resident and a friend of President Harding, Galbraith was well respected by all.

The rogue Kansas posts had no interest in following the dictates from Legion headquarters. The formation of the American Defense League of Kansas allowed Kitterman and others to continue to gather mobs and perform their cowardly attacks on Nonpartisan League people in the field.

Effie Mallon and the boys accompanied George to Kansas. The *Kansas City Kansan* of March 7, 1921 reported the visit of Mrs. George Mallon and son to the home of George's younger brother, A. E. Mallon, on Sandusky Avenue. The Medal of Honor citation printed in certain newspapers attributed the Mallon address as 1931 Hallock Street, Kansas City, Kansas. That was the address of Effie and the two boys while Captain Mallon was serving in the war in Europe. The Kansas address served the purpose of the Nonpartisan League by emphasizing Mallon's Kansas connection.

While Effie and the boys were in Kansas City, George returned to NPL headquarters in Minneapolis in early March. *Minneapolis Labor Review* reported Mallon as saying by early

March that the League was having success in Kansas, despite the opposition. While Captain Mallon was in Minneapolis, events were escalating in Kansas.

The Nonpartisan League in Kansas received numerous requests for former U.S. Senator J. R. Burton, a Nonpartisan League man in Salina, to speak in their communities. On March 12, Senator Burton and his wife, along with Professor George Wilson, were enroute to Ellinwood to address a gathering of some four hundred farmers. A mob intercepted the trio and forced them to turn around.

Carl O. Parsons and J. O. Stevic, editor of the *Kansas Leader*, set out with a sheriff's deputy from Ellinwood to search for their speakers.

Parsons and Stevic arrived at the county seat of Great Bend, where a mob in the courthouse halls threatened them. Law enforcement was present, but cowed in the presence of the unruly mob. Parsons and Stevic headed back toward Ellinwood. They saw that they were being followed before they realized they were being pursued by too many cars and trucks to count. The mob caught up with Parsons and Stevic not far out of Great Bend and forced them off their car off the road. The thugs pulled the two men out of their car, beat them badly and tar-and-feathered them. Parsons reported later that the mob was under the leadership of the local American Legion commander.

Minneapolis Labor Review published a taunting editorial in its March 18, 1921 issue under the headline: "Kansas in the Hoodlum Column". The editor pointed out:

> Captain George H. Mallon, in charge of the organization work for the League, was absent from the state at the time of the outrage. The residents of Kansas evidently believed it was the safest and most opportune time to carry out their hooliganism when Mallon was not at hand. There should be something stirring that mobs of yellow cowards don't care to participate if Mallon had

been present. Pershing's hero who met the enemy barehanded in the world war can be depended upon not to yield either quietly or gracefully to a coat of tar and feathers and the white-livered thugs that oppose the workers and farmers are well aware of that fact.

The editor charged the nonunion slush fund with paying for the devilment. He closed with:

> Kansas will be organized for the Nonpartisan League . . . It is not so surprising after all to see those who advocate slavery for the workers . . . resorting to mob violence to keep the farmers enslaved. Kansas is in the hoodlum column. It won't get out until the farmers and workers, completely organized in that state, clear its name, remove from office its lax officials and end the autocracy of greed in that state.

The *Kansas Leader* of March 21, 1921, bore a bold front-page headline, which read: "Mob Rule or Free Speech in Kansas? Which is It?" The question begged an answer.

Captain Mallon was incensed. He rushed back to Kansas and arranged for himself, Senator J. R. Burton and Professor George Wilson to speak in Marion, Kansas on March 16, 1921. Marion, the county seat, was fifty miles south of Junction City and Fort Riley. The captain was itching for a fight. No one knew what might happen. What seemed certain was that presence of Captain Mallon and a mob would be an explosive situation with violence sure to ensue.

The county sheriff of Marion, acting under orders from the mayor of Marion and the county attorney, forbad the NPL meeting, citing the potential for violence. Mallon fired off a telegram to Republican Governor Henry J. Allen. The Kansas Attorney General, under orders from the governor, then sent a telegram to Marion's mayor, the county attorney and county sheriff to report immediately to him at the state capitol in Topeka. Governor Allen made clear to the citizens of the state, "I am utterly out of sympathy with Townleyism ... preaching a class-minded doctrine ...". Nonetheless, the governor insisted that free speech was a right and that he wanted the facts before the people.

The issue of the opposition preventing Nonpartisan League organizers from speaking publicly was a clear violation of the Constitution. George Mallon published an angry letter to an opponent not long afterwards:

> Stop hiding behind the skirts of the American Legion and the American Defense League and come out like a man and debate this question. You hid behind these boys during the war and I hereby charge you with being a slacker during the great world war and I suppose that when your offspring ask you, "Where were you during the great World War, Grandpa?" you will answer and say — well, what will you say

Captain Mallon's blood was up. The mob action against his friends was unconscionable. He added to the letter:

> Will you tell them that you held that American soldier, who was carrying a banner charging you with being a slacker, while another man struck him? Don't deny this because I can prove it and if I ever get you on the platform I will do so before an audience.

Mallon and Burton next had their opportunity to speak in Topeka under the auspices of the Topeka Industrial Council. A large raucous audience filled the city auditorium and cheered wildly as Mallon and Burton threw jabs at the Marion city and county officials, but also at the governor and his attorney general.

The *Topeka State Journal* of April 19, 1921 reported on the remarks by the NPL leaders. Captain Mallon spoke first. He focused on his loyalty and the loyalty of the Nonpartisan League. His frustration and anger was evident. "It doesn't matter what I say here tonight. The kept press will come out tomorrow telling how I made a very disloyal speech." Mallon was in his usual awkward situation of attacking the American Legion, given his many friends and comrades in the Legion. Captain Mallon charged that, ". . . the capitalists put up the $275,000 necessary for the organization of the Legion, primarily for the purpose of using them as strikebreakers and they expect to get a hundred-fold out of their money." He attacked the chambers of commerce for their earnest efforts to stop the NPL. Captain Mallon also had harsh words for Governor Allen, who denounced the League on a regular basis. "Governor Allen doesn't believe in state ownership except for printing plants and cement plants and things which we can control himself."

Senator Burton referred to State Attorney General Hopkins as "Attorney General of the mob of the state of Kansas." Burton charged the governor as having "YMCA'd his way into the governor's office." Allen spent ten months with the Red Cross and six months with the YMCA. The Republican Party nominated Allen for governor while he was in France as head of communications for the Red Cross. Burton closed with an appeal to American Legionnaires. He urged them to understand that they had been "bamboozled." He reminded them that of the G. A. R., ". . . not a one of whom was ever a member of a mob which gathered to suppress the right of free speech."

Veterans, who were involved with the Nonpartisan League, were often also members of the American Legion. They understood well that Article II, Section 2 of the Constitution of the American Legion clearly stated, "The American Legion shall be absolutely non-political . . .".

In the midst of the political maneuvering and fighting, George Mallon suffered a personal setback in 1921. A doctor diagnosed him with chronic interstitial nephritis; i.e., serious kidney problems. Typical symptoms are fatigue, weight gain from water retention and elevated blood pressure. The end result is kidney failure. George Mallon was forty-four years old.

In the end, the Nonpartisan League effort in Kansas simply met too much resistance and began to lose ground. By August 1921, Captain Mallon became affiliated with Lincoln National Life Insurance Company as a salesman in Minneapolis with an office downtown in the Lincoln Bank Building.

Mallon did take part in the massive Labor Day parade in Minneapolis on September 5. Twenty thousand marched and many more lined the streets and cheered. George Mallon and Andrew G. Cooper addressed the crowd on behalf of the World War Veterans. This was among the final public appearances of Captain Mallon on behalf of the World War veterans.

George Mallon re-directed his involvement in veteran affairs in the fall of 1921, not long after the Labor Day event. The World War Veterans, once the primary competition for the American Legion for veteran membership, withered under attacks by the federal government, big business and rogue elements of the Legion for the organization's pro-labor stance.

The Veterans of Foreign Wars (VFW) had its origins in the soldiers returning from the Spanish American War. In the 1920s, the VFW was nowhere near the threat to the American Legion that the World War Veterans were. However, as the World War Veterans faded into obscurity, the Veterans of Foreign Wars began to rise to prominence in the 1920s.

Minnesota Veterans of Foreign Wars elected John Bowe of Canby as state commander in August 1921 at the annual encampment held at the Old Capitol in State Paul. The gathering also elected Dr. John Soper of Minneapolis as surgeon general and Colonel William H. Donahue of Minneapolis as judge advocate. All three were men whom Mallon would come to know well in the post-war period. Bowe announced that George H. Mallon would be his chief of staff for the Department of Minnesota. Under the leadership of Bowe and Captain Mallon, the Minnesota VFW greatly broadened its activities.

Bowe, like Mallon, served in both the Philippine War and in the World War in France. Both Bowe and Mallon were recipients of the France's Croix de Guerre. That was where the similarity ended. Bowe was English, an immigrant to the U.S. as a young man. He volunteered for the 13th Minnesota Volunteer Infantry Regiment and saw considerable action in the Philippine War as a private. When the Great War broke out in Europe in 1914, Bowe did not wait for the United States to enter the conflict. He joined the French Foreign Legion and fought in the trenches. He was a man of considerable passion and somewhat of an eccentric.

CHAPTER THIRTEEN
DEDICATION OF
VICTORY MEMORIAL DRIVE

Earlier in 1921, while Mallon was still active with the Nonpartisan League, a major event took place in Minneapolis to honor those from Hennepin County, who gave their lives in the World War. Mallon took leave of the action in Kansas and returned to Minneapolis for the event.

There was a movement across the nation to commemorate those who gave their lives in the world war in Europe. Among the most famous was the Liberty Memorial in Kansas City. Charles Loring, Minneapolis Parks Commissioner, and Theodore Wirth, Parks Superintendent, conceived of a unique monument to the 568 men and women of Hennepin County, who made the ultimate sacrifice. The result was the transformation of nearly four miles of the Glenwood-Camden Parkway between Lowry Avenue and Camden Park in residential north Minneapolis into a natural monument. Elm trees lined either side of the boulevard, each tree a memorial to one individual. Next to each tree was a wooden cross honoring that individual. What became Victory Memorial Drive runs approximately six-tenths of a mile from east to west then turns south and extends for a mile from north to south. An impressive stone monument with a flagpole stood at the inflection point.

The dedication was under the auspices of the American Legion. David L. Sutherland, the commander of Minnesota American Legion, chaired the General Committee. Newly elected Lieutenant Governor Louis L. Collins, also a Legionnaire, chaired the Speakers Committee.

The original speaker was to have been Frederick W. Galbraith, Jr., the national commander of the American Legion.

He was a passenger in a car and died in an accident in Indianapolis on June 9, just two days before the dedication. Everyone felt the tragic loss of such a fine man. The Legion set up a draped chair with a wreath on the stage in his honor. Collins secured the services of three Legionnaires to speak in Galbraith's place. These were Congressman Royal C. Johnson of South Dakota, Hanford MacNider of Mason City, Iowa and John J. Emery of Indianapolis, Vice Commander of the Legion.

The dedication of the monument took place on June 11, 1921. With the war still a fresh and painful memory, the dedication was a major event in the city. More than thirty thousand people lined the route to pay respects.

Colonel William H. Donahue, a good friend of George Mallon's and a fellow Irishman, was grand marshal of the first division. Donahue served under Leach in the 151st Field Artillery during the war. The first division started at the south end at 33rd Avenue North and moved from south to north up Victory Memorial Drive. The First Minnesota Infantry Band led a long line of American Legion posts.

Colonel E. E. Watson was grand marshal for the second division. This column started on the east end from Humboldt Avenue and marched westward along Victory Memorial Drive. George Mallon, as a marshal, led the World War Veterans. Ole G. Sanstad, the first state commander of the Veterans of Foreign Wars, led that contingent. Sanstad, like Mallon, was a veteran of the Philippines War.

A salvo of 48 guns fired by the 151st Field Artillery, commanded by Colonel George Leach, began the formal dedication. The mayoral election happened to be in two days. George E. Leach was facing Socialist Thomas Van Lear, who lost his re-election bid in 1919. The headline of the June 12 issue of the *Minneapolis Sunday Tribune* read: "Voters Wait for Opening of the Polls to Finish War Against Townleyism. George Leach, Ex-Soldier, Asks Election Over Radical Socialist." Leach was a member of both the American Legion and the Veterans of Foreign War.

Mallon was friends with both men, although closer to Van Lear. Van Lear served four years in the Army, including re-enlisting in Chicago in 1898 when the Spanish American War began.

When the two columns met at the stone memorial at the bend in Victory Memorial Drive, the prepared remarks commenced. Judge Eli Torrance delivered the dedicatory address. He was a Civil War veteran and past National Commander-in-Chief of the Grand Army of the Republic. Judge Torrance's message included: "And to you, the members of the American Legion who constitute the new Grand Army of the Republic, the old Grand Army of the Republic extends its most cordial and fraternal greetings. We recognize you as the uncompromising champions of a world democracy and the invincible defenders of American institutions."

The *Minneapolis Morning Tribune* on the following morning reported:

> Lieut. Gov. Louis L. Collins and Capt. George H. Mallon, who have been opponents in political contests, but are warm friends, met at the speakers' stand and almost embraced.

The paper included a photograph of the two soldiers smiling and talking. The photograph said much about reconciliation in the aftermath of the hard-fought election in November. Their bond of soldier brotherhood overrode any political differences. The caption read:

> Captain George H. Mallon, one of Pershing's hundred heroes of the war, and Lieutenant Governor Louis L. Collins, little corporal of the 151st, held an informal reunion of their own at the Memorial Drive dedicatory ceremonies yesterday. Having met by chance, the two threw arms around each other in greeting, then shook hands as the photographers got into action.

Crowd of thousands gathered
Dedication of Victory Memorial Drive
Minneapolis, June 11, 1921
Courtesy of Hennepin County Library

Hennepin County Legionnaire posted the same photo and caption on page one of its June 16, 1921 issue. The image and the caption said much about veteran solidarity and camaraderie, even in a time when their organizations were involved in intense competition.

Two days later, George Leach defeated George Mallon's close friend Thomas Van Lear in the Minneapolis mayoral election. Van Lear's defeat was yet another blow to the Nonpartisan League, which was hoping to recover from the defeat in the previous November of its state slate, including Mallon for lieutenant governor.

On June 18, just one week after the dedication of Victory Memorial Drive, General John J. Pershing, newly appointed Chief of Staff of the United States Army, gave an exclusive interview to a news agency. He wanted to speak out for Citizens' Military Training Camps. Pershing made clear that he was speaking as an American citizen, not as a soldier, when he said the camps would become the foundation of the nation's military policy.

George Mallon & Louis Collins post-election
from *Hennepin County Legionnaire*, June 16, 1921
Courtesy of Al Zdon, American Legion headquarters

    The National Defense Act was enacted on June 4, 1920. The legislation established the U. S. Army to be comprised of three parts: the Regular Army, the National Guard and the Army Reserve. As a compromise to the defeat of Universal Military Training, the act authorized Citizens' Military Training Camps to be held throughout the United States each summer (this practice continued to 1940). Unlike the National Guard and the Army Reserve, citizens could attend volunteer, month-long training camps at U.S. military posts without making a commitment and without any obligation to respond to being called up for active duty. Volunteers paid their own expenses.

    On August 10, 1921, General Pershing visited the Twin Cities for his first time since the end of the war. The primary reason for the visit was to inspect the Citizens' Military Training Camp at Fort Snelling. Minnesota Governor J. A. O. Preus and Minneapolis Mayor George Leach escorted General Pershing on what the *Minneapolis Morning Tribune* of the following day termed a "Rush Event." Pershing attended a breakfast at the Minnesota Club in St. Paul, addressed the Veterans of Foreign Wars at the Old Capitol, spoke to members of the National Guard at the St. Paul Armory and gave a speech at a luncheon at the St. Paul Athletic Club. General Pershing began his afternoon at Fort Snelling. He inspected the Citizens' Military Training Talk and spoke to the participants.

The Minneapolis VA Medical Center was not yet a reality. In 1921, the U.S. Veterans Bureau leased the Aberdeen Hotel in St. Paul and the Asbury Hospital in Minneapolis to provide for the hospitalization of disabled war veterans. General Pershing went to both facilities and visited with his former soldiers of the A.E.F. Afterwards, the general rode the length of Victory Memorial Drive in an open touring car, accompanied by Mayor Leach, Theodore Wirth and Donald Sutherland, the chairman of the American Legion Victory Memorial Committee. A photo of the group appeared in the next morning's *Tribune*.

There is no indication that General Pershing met with George Mallon while in the Twin Cities. It is possible that Mallon was present at the Old Capitol when Pershing addressed the Veterans of Foreign Wars — Mallon being a member. No mention was made of his presence. It is likely that Pershing snubbed Mallon on his visit. Pershing facilitated the formation of the American Legion and was widely considered to be one of the Legion's staunchest allies. Mallon was the foremost member of the World War Veterans, which competed with the Legion for membership, as well as attacked some of its reactionary policies. Pershing was adamant about the necessity of Universal Military Training. Mallon was at the other end of the spectrum; strongly opposed to Universal Military Training. There is an irony in that Mallon's most commonly repeated honorific was "one of Pershing's Hundred Heroes." Of course, the Hundred Heroes was essentially a device to facilitate the sale of Victory Liberty Loans. Nonetheless, Mallon was a Medal of Honor recipient and Pershing himself pinned the medal on Mallon. There is no record of them every talking after the Medal of Honor ceremony at Chaumont, France.

The natural monument known as Victory Memorial Drive has stood the test of time, even if it is unknown to many to whom World War One means little. The initial trees, Moline elms, were unable to endure the frigid winter. The Parks Department replaced them in 1925. The initial wooden crosses did not last. The Parks Department replaced them in 1928 with bronze crosses and stars of David set into the ground at the base of each tree. Of course, the plaques are only visible to those who park their cars and walk up to the trees. Victory Memorial Drive to this day remains a reminder of the cost of war and the ultimate sacrifice of some for democracy.

On May 25, 1930, the Grand Army of the Republic dedicated a large bronze statue to Abraham Lincoln on the boulevard in memory to those who gave their lives in the Civil War. Minnesota Civil War veterans in attendance included north Minneapolis resident Henry Mack, an escaped slave who served in the U.S. Colored Troops during the war.

Magnificent elm trees adorned the avenue during the 1950s, creating a serene green canopy under which cars cruised. The Dutch Elm disease, which ravaged the continent in the 1960s and 1970s, did the same to Victory Memorial Drive. The Parks Department replaced each elm tree with a hackberry tree, which today stand silent testament to the dead.

CHAPTER FOURTEEN
TOMB OF THE UNKNOWN SOLDIER

Just as the June 1921 dedication of Victory Memorial Drive honored those from Minneapolis and Hennepin County, who gave their lives in The World War, a much larger event took place in Washington, D.C. in November 1921 to honor the many unidentified American soldiers, who died in France in The World War.

On March 4, 1921, the United States Congress approved a measure to return the body of an unknown American soldier from France and inter the body in a special tomb at the Memorial Amphitheater in Arlington National Cemetery. The installation was to take place on Armistice Day, November 11, 1921.

The United States Government invited George H. Mallon, by virtue of his being a Medal of Honor recipient, to attend the ceremony at the new Tomb of the Unknown Soldier in Arlington National Cemetery. The government paid all his expenses. The official representatives of the State of Minnesota to the event were Minneapolis Mayor George E. Leach and Lt. Governor Louis L. Collins, both Republicans and veterans of the 151st Field Artillery. Mallon attended as a decorated soldier of the nation. All three belonged to the Veterans of Foreign Wars, as did Captain Mallon.

The major Minneapolis newspapers only mentioned that George Leach, Louis Collins and Colonel Bjornstad would attend the dedication of the Tomb of the Unknown Soldier in Washington, D.C. The three veterans were the official representatives of the State of Minnesota. There was no mention of Captain George Mallon, who was invited by virtue of being a Medal of Honor recipient. *Minneapolis Labor Review* briefly mentioned Mallon's invitation in just a few sentences.

In October 1921, personnel at each of the four American cemeteries in France exhumed the body of a member of the American Expeditionary Force, who died in combat and whose identity was unknown. One each from Aisne-Marne, Somme, St. Mihiel and Meuse-Argonne. Each body was placed in an identical casket and an identical shipping case. Trucks delivered the remains to the city hall of Chalons-sur-Marne. The hall was decorated with French and American flags. French officials joined the American delegation at City Hall. Soldiers on duty set each casket on top of its shipping case. They then re-arranged the caskets and placed in different cases. A sergeant selected the shipping case, which contained the remains which would become the Unknown Soldier. No one knew from which cemetery the remains came.

A train transported the remains of the Unknown Soldier to Paris overnight and then to the port of Le Havre. Army bearers handed the casket to Navy and Marine bearers, who carried the casket on board the cruiser USS Olympia, Admiral Dewey's flagship in the Spanish American War. The new destroyer USS Rueben James (which the German U-boat sank in October 1941, prior to the declaration of war) escorted the Olympia. Eight French ships accompanied the two American ships to international waters. The Olympia sailed up the Potomac and docked at the Washington Navy Yard on November 9. A military escort then took the casket bearing the Unknown Soldier to the Capitol, where he lay in state in the rotunda until November 11.

On the morning of November 11, 1921, Armistice Day, eight body bearers, followed by twelve honorary pallbearers, carried the casket down the steps of the Capitol to the horse-drawn gun carriage. Among the honorary pallbearers were Lt. Colonel Charles Whittlesey, Sergeant Alvin York, First Sergeant Sam Dreben and First Sergeant George Wanton. Whittlesey, survivor of the Lost Battalion, would commit suicide two weeks later. Wanton, a Buffalo Soldier of the 10th Cavalry, was a Medal of Honor Recipient of the Spanish American War in Cuba. Dreben was an acquaintance of George Mallon's from his membership in the World War Veterans.

George H. Mallon (far left) and other Medal of Honor recipients at the dedication of the Tomb of the Unknown Soldier
Courtesy of George Mallon

Motorcycle police, followed by mounted officers, led the vast procession down Pennsylvania Avenue across the Potomac River to Arlington National Cemetery. Brigadier General Harry Bandholtz, commander of the Military District of Washington, who planned the ceremony, was next, mounted on horseback and accompanied by his staff. A military band was next, playing a funeral dirge, its cadence marked by muffled drums. Then soldiers and sailors, including a platoon of infantrymen with fixed bayonets and horse-drawn machine guns. Four clergymen immediately preceded the horse-drawn caisson bearing the flag-draped casket. The honorary pallbearers walked alongside on either side of the caisson. Six black horses with rigid uniformed riders pulled the caisson.

There was no cheering or flag waving. The vast crowd was silent. The only sounds were the muffled drums and the horses' hoofs and boots of the soldiers on the pavement.

President Harding and General Pershing followed the caisson, along with leading statesmen & jurists and Pershing's his aides, including George Marshall. A long line of cabinets members, governors, senators & congressmen, and high-ranking military officers were next.

The Medal of Honor recipients, including Captain Mallon, marched eight abreast behind the dignitaries. Behind them were 132 selected World War veterans from the various states, including Collins and Leach of Minnesota.

November 11, 1921 was a major event in the history of the United States of America. George Mallon was present, not as an observer, but as a participant. President Warren Harding's eulogy, "There must be, there shall be, the commanding voice of a conscious civilization against armed warfare," surely resonated with Captain George Mallon. With the closing of the dedication of the Tomb of the Unknown Soldier, Mallon gathered with his comrades from the World War Veterans organization for one more task at hand.

Socialist leader and pacifist Eugene V. Debs had been in

Atlanta Prison since April 1919. Despite his incarceration, Debs ran for president in the 1920 election against the Republican candidate Warren G. Harding and the Democratic candidate James M. Cox. Inmate 9653 received 917,799 votes, 3.4% of those cast. Harding won the presidential election by a landslide. His momentum helped the Republican ticket of Preus and Collins defeat the Nonpartisan League slate of Shipstead and Mallon in the Minnesota gubernatorial race.

The immediate task before newly inaugurated President Harding in March 1921 was to restore the nation to post-war normalcy. A very important matter before the nation was the incarceration of Eugene V. Debs. President Woodrow Wilson refused to pardon Debs or commute his sentence before leaving office and continued to hold steadfast on that position. The American Legion took a very strong position that Debs remain in prison for the term of his sentence. *The New York Times* of March 26 noted, "He deserved ten years' imprisonment if any man ever deserved it." Even the Harding's wife was opposed to him releasing Debs.

Minnesota Leader of July 2, 1921 announced, "World War Veterans Plan Unemployment Protest to Harding". Two thousand members of the World War Veterans met in Chicago for a national convention. There was discussion among the World War Veterans about adjourning and reconvening in Washington, D.C. to protest on the steps of the White House. A major issue was the unemployment of 1.5 million ex-servicemen. The veterans wanted to protest attempts to deprive the state of North Dakota of its elected government, which the Nonpartisan League controlled. The protesters also wanted to demand that President Harding release Eugene Debs from prison. The hugely popular Debs was among 140 men and women, whom President Woodrow Wilson, Attorney General A. Mitchell Palmer and young J. Edgar Hoover placed in prison for opposition to the war.

Eugene V. Debs of Terre Haute, Indiana, was one of the most well-known people in the United States. The extremely charismatic speaker addressed a crowd in Canton, Ohio on June 6, 1918. The federal government charged Debs with violating the Sedition Act.

Debs opened his Canton speech by saying:

> . . . it is extremely dangerous to exercise the constitutional right of free speech in a country fighting to make democracy safe in the world. I realize that, in speaking to you this afternoon, there are certain limitations placed upon the right of free speech. I must be exceedingly careful, prudent, as to what I say, and even more careful and prudent as to how I say it. I may not be able to say all I think; but I am not going to say anything that I do not think.

And near the end of his speech, Debs said:

> The master class has always declared the wars; the subject class has always fought the battles. The master class has had all to gain and nothing to lose, while the subject class has had nothing to gain and all to lose –especially their lives . . .
> And now for all of us to do our duty! The clarion call is ringing in our ears and we cannot falter without being convicted of treason to ourselves and to our great cause. Do not worry over the charge of treason to your masters, but be concerned about the treason that involves yourselves. Be true to yourself and you cannot be a traitor to any good cause on earth.

Despite Debs' careful wording, the Court found intent in his general critique and convicted Debs of sedition. The judge sentenced him on November 18, 1919 — more than one year after the end of the war — to ten years in prison. The

Sedition Act was not only unconstitutional, but the war was over. Nonetheless, the United States Supreme Court upheld the conviction on Mach 10, 1919. The American legal system failed to protect the constitutional right of Eugene Debs to speak, because the government feared the power of the brilliant orator.

Debs was a man of fierce conviction, very much like George Mallon. Debs' reference to the brotherhood of man, no doubt, appealed to Mallon. They both opposed militarism and sought to improve the working conditions and lives of the working class. Both were well liked and well-spoken men, although Debs was the more charismatic. Both were warriors, each in his own way.

In the aftermath of the service at Arlington National Cemetery, Captain Mallon led a group of World War Veterans to the White House. Theirs was not just any group, but a clique of highly decorated war heroes, whose patriotism could not be doubted. The World War Veterans had one simple mission; i.e., to appeal to President Harding to release Eugene V. Debs and others similarly imprisoned for opposing the war.

The Bureau of Investigation (predecessor of the FBI) closely followed the activities of the World War Veterans. Director William J. Burns considered WWV to be extremely radical. He warned President Warren Harding of the intent of the World War Veterans to present him with a petition to release war protestors from prison.

George Mallon understood well the value of his Medal of Honor for allowing him credibility as to his patriotism and loyalty. He gathered around himself a cadre of similarly decorated men. They were as follows:

Berger Lohman of Chicago served as private with the 132nd Regiment/33rd Division. This was the same regiment in which Mallon commanded Company E. Private Lohman was decorated for heroism in the advance across the Meuse on Oct. 9, 1918, one week after Captain Mallon was wounded. His Medal of Honor citation read:

When his company had reached a point within 100 yards of its objective, to which it was advancing under terrific machine gun fire, Pvt. Lohman voluntarily and unaided made his way forward after all others had taken shelter from the direct fire of an enemy machine gun. He crawled to a flank position of the gun and, after killing or capturing the entire crew, turned the machine gun on the retreating enemy.

Clayton K. Slack of Wisconsin served as a private in the 124th Machine Gun Battalion/33rd Division. This was the same division in which Mallon and Lohamn served. He was decorated for heroism in the advance across the Meuse on Oct. 8, 1918. His Medal of Honor citation read:

Observing German soldiers under cover 50 yards away on the left flank, Pvt. Slack, upon his own initiative, rushed them with his rifle and, single-handed, captured 10 prisoners and 2 heavy-type machineguns, thus saving his company and neighboring organizations from heavy casualties.

Private John Joseph Kelly of Chicago was a runner in 78th Company of 6th Regiment Marines, 2nd Division. He fought at Belleau Wood and took part in the St. Mihiel Offensive. The fighting for heavily fortified Blanc Mont Ridge in October 1918 was as bloody as anything the Devil Dogs faced. Kelly was decorated for heroism in the initial assault on Oct. 3, 1918. His Medal of Honor citation read:

For conspicuous gallantry and intrepidity above and beyond the call of duty in action with the enemy at Blanc Mont Ridge, France, October 3, 1918. Private Kelly ran through our own barrage one hundred yards in advance of the front line and attacked an enemy machine-gun nest, killing the gunner with a grenade, shooting another member of the crew with his pistol and returned through the barrage with eight prisoners.

When he reappeared through the smoke and haze, Kelly is said to have shouted, "Just what I told you I'd do!" While the Marines took the ridge in their initial assault, the supporting units failed to cover their flanks, leaving them isolated and exposed to counterattacks from all directions. The battle ended on October 27. Kelly also received the Silver Star with four oak leaf clusters, the Purple Heart and the Croix de Guerre, among others.

Colonel Nelson Miles Holderman, as a captain in 1918, commanded Company K, 305th Regiment, 77th Division during the Meuse-Argonne Offensive. Holderman's company was the only relieving unit to reach Major Charles Whittlesey's Lost Battalion (Whittlesey actually had two battalions). With no other relief, Company K became part of the Lost Battalion with Holderman given responsibility for holding the right flank. Despite being severely wounded, Holderman continued to lead the defense. For his unflinching courage and leadership, he received the Medal of Honor. His Medal of Honor citation read:

> Captain Holderman commanded a company of a battalion which was cut off and surrounded by the enemy. He was wounded on 4, 5, and 7 October, but throughout the entire period, suffering great pain and subjected to fire of every character, he continued personally to lead and encourage the officers and men under his command with unflinching courage and with distinguished success. On 6 October, in a wounded condition, he rushed through enemy machinegun and shell fire and carried 2 wounded men to a place of safety.

Holderman also received the Silver Star, the Purple Heart and the Croix de Guerre.

Sam Dreben was the most colorful figure of the bunch. While he was not a Medal of Honor recipient, General Pershing was quoted as referring to Dreben as "the finest soldier and one of the bravest men I ever knew." He was legendary for being fearless and reckless, as well as for his many exploits and adventures. A Russian Jew, Dreben enlisted in the U.S. Army just a few months after arriving in the United States. Like Mallon, he served in the Regular Army in the Philippines War. Also like Mallon, he earned a promotion to sergeant. Dreben subsequently fought as a mercenary in Central America and Mexico. He was a machine-gunner for Francisco Madero during the Mexican Revolution and, later, smuggled arms to Pancho Villa. After Villa raided New Mexico, Dreben scouted for Pershing's Punitive Expedition in search of Villa. When the U.S. entered the world war, Dreben enlisted and served as a sergeant in the 141st Infantry Regiment, 36th Division. He received the Distinguished Service Cross for his valor at St. Etienne in October 1918. His citation read:

> . . . while serving with Company A, 141st Infantry Regiment, 36th Division, A.E.F., near St. Etienne, France, 8 October 1918. Sergeant Dreben discovered a party of German troops going to the support of a machine-gun nest situated in a pocket near where the French and American lines joined. Sergeant Dreben called for volunteers and, with the aid of about 30 men, rushed the German positions, captured four machine-guns, killed more than 40 of the enemy, captured two, and returned to our lines without the loss of a man.

It was at the head of this distinguished band of celebrated American heroes that Captain Mallon respectfully approached the White House to request an audience with President Harding. Mallon learned from the moment he returned home from the war that his Medal of Honor allowed him a platform from which to speak and a level of credibility from those who challenged the loyalty and patriotism of anyone with whom they disagreed. He knew that the war heroes, whom he gathered around himself, carried far more weight with the public, as well as with the president, than just George H. Mallon, one Medal of Honor recipient. All were members of the "radical" World War Veterans. No one, even the most bitter ultra-conservative, dared question the patriotism of this distinguished group of gallant men. Newspapers carried photos of the veterans at the White House. Three leaders of the World War Veterans, who were not war heroes, but were veterans, were also in the photo. Andrew G. Cooper of Minneapolis was national chairman. John M. Levitt was eastern division chairman and Carl Parsons was Minnesota state chairman.

President Harding did meet with Captain Mallon and his cadre. There was no photo op, no doubt, due to the sensitive nature of the meeting. The three leaders of the World War Veterans presented their letter to the president. The letter, which the Medal of Honor recipients presented to President Harding on November 13, 1921, read:

> We, the undersigned, holders of the Congressional Medal of Honor, wish at this time to second the memorial tendered you by the World War Veterans, a copy of which is attached, believing that the sentiments they express represent the view of the rank-and-file of ex-service men the country over. Had the comrade whom we honored on Armistice Day returned to America alive, he would perhaps be appealing with us to you for the release of these prisoners.
>
> You said at this bier on Armistice Day: "His patriotism was none less if he craved more than triumph of country; rather, it was greater if he

hoped for a victory of all human kind. Indeed, I revere that citizen whose confidence in the righteousness of his country inspired belief that its triumph is the victory of humanity.

Mr. President, it is that very kind of citizen whom the government is today holding behind prison bars for loyalty to their ideals. It was no easy task for them to risk unpopularity and prison to maintain these ideals against the majority of the people in the time of war. Their loyalty to the interest of humanity as a whole, even against their country's decision to join in the war, was what moved them to express the opinions which sent them to prison. We disagree with the methods of the men in prison. We followed, ourselves, the opposite course. But we respect them for their opinions and their courage, as we respect our own comrades.

We understand that every country in the world, which engaged in the great war, has long since released from prison those who, like these prisoners, opposed the war. May we ask, Mr. President, why America with her democratic ideals, should wait so long to do an act of justice and goodwill?

Again, on Armistice Day, you said: "I can sense the prayer of our people, of all people, that the Armistice Day shall mark the beginning of a new and lasting era of peace on earth, goodwill among men.

May we ask as ex-soldiers, holders of the Congressional Medal of Honor, that you "mark the beginning of this new era" by expressing in an amnesty the generosity and goodwill of those of us who fought. As Americans, we seek a return to that condition of goodwill and love of neighbor, which obtained throughout our beloved country before the war.

Faithfully yours,

Clayton K. Slack
Captain George H. Mallon
John J. Kelly
Berger Lohman

Holderman and Dreben, who accompanied their comrades to the White House, were not signatories to the previously prepared letter to the president.

In addition to being the third anniversary of the end of hostilities of The World War and also the dedication of The Tomb of The Unknown Soldier, November 11, 1921 had additional importance. The United States was not party to the Treaty of Versailles, signed on June 28, 1919. President Woodrow Wilson opposed the harsh terms, which Britain and France imposed on Germany. That was contrary to his Fourteen Points. The United States and German finally signed a peace treaty on August 25, 1921. After ratifications, the treaty became effective on November 11, 1921. As a result, the United States remained at war with Germany until November 11, 1921, the date on which Captain Mallon and his Medal of Honor comrades met with President Harding. It was only at this point with the war officially over that the president would seriously consider releasing Debs from prison and putting The World War behind the United States. In addition, it was appropriate that the first arms control conference in history was set to open in Washington, D.C. on the following day, November 12. President Harding called the Washington Disarmament Conference, also known as the Washington Naval Conference, and invited the world powers.

The New York Times of November 15, 1921 covered the

story, opening with "Two legations of ex-service men, one consisting of four wearers of the Congressional Medal and the other of officials of the World War Veterans, today asked President Harding to pardon all political prisoners." The Times printed both petitions – that of the Medal of Honor recipients and of the World War Veterans and clearly stated, "the four Congressional Medal of Honor men, who were all members of the World War Veterans". The article also stated, "The memorial of the Congressional Medal of Honor men was presented by Captain George H. Mallon … and A.G. Cooper, National Chairman, of Minneapolis, handed the president the petition of the World War Veterans.

The November issue of Chicago publication *Debs Freedom Monthly* included a story titled, "Heroes Plea For Liberty."

> Former servicemen are taking a most active part in the movement for the release of all political prisoners. The resolutions presented to the President of the United States recently are so lofty and inspiring and reflect so clearly the nobility of the boys in the ranks of the World War Veterans that we are glad to publish them in their entirety. When the final history of freedom is written in burnished rows of steel the action of this organization of unknown heroes will be given the full credit it so fittingly deserves.

The *Indianapolis News* on November 28 published a telegram, which the three World War Veterans officers sent to Theodore Debs, the brother of Eugene V. Debs:

> We, the undersigned officers of the World War Veterans, who are here in Washington to honor our unknown comrade, greet you as the best known of a band of 140 political prisoners in federal prisons for opposing the war. You were inspired by the

same ideals as we who fought. We differed only in the means of achieving those ideals. We respect your views and your courage and we demand for you the same freedom of opinion which we have enjoyed ourselves and, believing that, we have appealed to the President to grant a general amnesty of all such prisoners.

Our appeal to the President has been indorsed in a memorial by holders of the Congressional Medal of Honor. We assure you that the hearts of the men, who fought for American ideals, are with you. Your release and that of other prisoners will mark the recognition of those principles on which our country's institutions rest.

The American Civil Liberties Union (ACLU) issued a press release to announce a joint effort with several other groups under the name Joint Amnesty Committee to secure the release of all federal political prisoners.

The attitude of the administration has been favorable to the release of individual prisoners "on their merits," but practically none have been released, presumably because of the opposition of the American Legion.

The well-known journalist Gilson Gardner, Washington correspondent for the Scripps newspaper syndicate, headed the effort. The committee also included activist Belle Case LaFollette, wife of Progressive U.S. Senator "Fighting Bob" LaFollette and Rev. John A. Ryan, a professor at Catholic University in Washington, D.C. and a leading Catholic activist for social justice. The press release stated, "Among the organizations active in this amnesty work are the World War Veterans. The statement commented on the meeting of the four Medal of Honor recipients, led by Captain Mallon, with the president.

Minneapolis Labor Review ran a story on November 11 and re-ran the same piece on page one a week later. The headline read, "Medal of Honor Men Picketing Harding." The article began:

> Led by congressional medal of honor men, the World War Veterans are picketing the disarmament conference and the White House for the release of Debs and all class war prisoners. Heading the medal men is George H. Mallon of Minneapolis ... This is the most effective and spectacular work for the release of the class war prisoners that has yet been attempted ... Harding, it is believed, will be forced to yield.

It is not known when George Mallon returned to Minneapolis from Washington, D.C. However, he responded to the articles in *Minneapolis Labor Review* and elsewhere on December 22. As major newspapers often did not cover Mallon and *Minneapolis Labor Review* might not have been inclined to publish Mallon's statement, the only known paper to publish Mallon's response was *The Labor World* in Duluth on December 24, 1921. The headline read, "Mallon Scorns Parley Pickets, War Hero Declares No Surer Way of Keeping Debs in Prison Could Be Devised."

> Picketing of the disarmament conference by the World War Veterans and other liberals anxious to obtain the release from prison of political prisoners was declared today by Captain George H. Mallon to be a mistake in tactics. He denied published reports from Washington he had participated in the demonstrations with other World War Veterans.

"No surer way of keeping Eugene V. Debs and other politicals in jail could be devised than the senseless picketing," Captain Mallon declared. "With other congressional medal of honor men, I presented a memorial to President Harding, urging release of political prisoners. The memorial was dignified and in an interview with the president I felt he was impressed with the injustice done political prisoners by keeping them in jail after the law under which they were convicted had been repealed. I feel anything we were able to accomplish has been undone, however, by the picketing. Demonstrations of that kind invite opposition and are calculated to arouse a feeling of resentment and obstinacy.

Captain Mallon went to Washington to attend the burial of the Unknown Soldier on Armistice Day as the federal government's guest.

By the time, the *Labor World* article came out, there was bigger news. The president was responding to the clamor to release Debs from prison.

President Harding did not respond immediately to the request by Mallon and his decorated comrades. Six weeks after meeting Mallon, President Harding announced his decision. On December 23, 1921, Harding commuted Eugene Deb's sentence, effective Christmas Day. He declared it was time for the nation to move past the war. Harding carefully chose to not pardon Debs in an effort to balance public reaction to his decision. The White House released a statement:

> There is no question of his guilt and he actively and purposely obstructed the draft. In fact, he admitted it at his trial, but sought to justify his

action. He was by no means, however, as rabid and outspoken in his expressions as many others and, but for his prominence and the resultant far-reaching effect of his words, very probably might not have received the sentence he did.

In the midst of heated opposition, the president insisted that he was fulfilling his responsibility to return the nation to normalcy. His was an act of immense moral courage.

President Harding, at the same time, also commuted the sentence of twenty-three others imprisoned for opposition to the war. *The New York Times* reported that less than a third of those released were I.W.W. members (Wobblies), but that none of them advocated sabotage or force. Harding continued to review cases and release other political prisoners in the following months. Harding died suddenly in August 1923. Debs' health never recovered from his time in prison. He died of heart failure at his home in 1926.

There is no record of the reaction of George Mallon as to the president's move to release Debs. George Mallon simply returned home to his many pursuits. He spoke to student cadets at Minneapolis South High School on January 11, 1922.

After considerable bloodshed, the Irish War of Independence ended with the signing of the Anglo-Irish Treaty on December 6, 1921. The agreement provided for the establishment of The Irish Free State, consisting of 26 of Ireland's 32 counties, within the year. Irish Americans lifted a toast. Six northern counties remained in the United Kingdom as Northern Ireland. One of those was County Tyrone, George Mallon's ancestral homeland. His feelings must have been mixed. The Irish Free State came into being on December 6, 1922 without the Mallons of County Tyrone. Violence continued in County Tyrone for decades, well beyond George Mallon's lifetime.

Medal of Honor recipients, World War Veterans at White House, November 11, 1921
Left to right: Unidentified, Lohman, Kelly, Holderman, Mallon, Slack, Dreben, three unidentified
Courtesy of Library of Congress

U. S. Congressman Thomas Schall was one of ten men representing Minnesota in the 67th Congress. All were Republican. Schall was unsuccessful in convincing Captain Mallon to attend a session so that Schall and the rest of Congress could honor him. Seven months later, in July 1922, Congressman Schall praised George Mallon in an address before Congress.

> Captain Mallon was a delegate to the burial of the unknown dead. I tried my best to get him into the gallery that I might have the honor to proudly introduce him to the house, but his retiring, simple and modest nature would not cooperate with me.

Schall was legally blind, the result of an accident more than a decade earlier. He had a full-time page to assist him. Congressman Schall took the podium on the occasion of the visit of ninety blind veterans to Congress.

> My friend, my constituent, my neighbor, Captain George Mallon, has been officially designated one of the 100 heroes. All our allies have honored him with distinguished notice of exceptional bravery. He well typifies the ordinary American soldier, who performed his duty without boast or consciousness that he was doing anything out of the ordinary.
>
> Up to the present time, though I have tried and tried, I have been unable to get him to talk of his war experiences. George is a big-hearted, big-fisted, big-bodied, six-footer, gentle, simple, modest, intelligent, who has earned his living by the use of his hands and the sweat of his brow – a plain American who has had no advantages except those of his own making . . . his character and war record aptly illustrate the soul of the plain American soldier who has not had the approving

stamp of official notice, who makes no claims as to heroism, who performs his duty without thought of reward or praise … I wanted to tell this house of the war department's citation of this steamfitter soldier, whose heroism blazes forth even from the cold meager details of officialdom.

Schall went on to serve as U.S. Senator from Minnesota from 1925 until his death in 1935.

CHAPTER FIFTEEN
HENNEPIN COUNTY COMMISSIONER

President Warren G. Harding carried other war-related burdens on his shoulders, besides the matter of Eugene Debs. In January 1920, in the aftermath of The World War, the United States economy sank into a deep deflationary recession. The adjustment back to a peacetime economy was painful. When Harding took office in March 1921, the nation remained immersed in the throes of an economic slowdown.

The federal government financed U.S. involvement through four Liberty Loan drives during the war and the fifth drive, the Victory Loan, after the war. The government's colossal advertising campaign contributed greatly to the program's success. That included the memorialization of Captain George Mallon's fist into the jaw of the German officer. All five of the drives were oversubscribed. The government borrowed $21.5 billion in all. In contrast, the national debt was just over $1 billion at the end of 1916 on the eve of the United States' entry into the Great War in Europe.

The debt led to a contraction of credit. Expectations of further deflation caused investment and production to decline. Wartime production to meet the needs for the war drastically distorted the private market. Two million soldiers returned from Europe and re-entered the job market along with another three million in the service, who did not go overseas. That and declining production led to unemployment on a vast scale and to wage stagnation for those fortunate enough to have jobs. Agricultural prices for American crops fell as European crops returned to the market.

The Federal Reserve Bank did not respond to the crisis by adjusting interest rates. Harding slashed the federal budget, increased the tax basis and cut taxes. The recession ended in mid-summer 1921.

George H. Mallon
Courtesy of Diane Gossage

The Roaring Twenties was a prosperous era. Minneapolis was a wide-open town. Never mind that Prohibition went into effect in 1920 and made the manufacture, transportation and sale of alcohol illegal. That allowed many small-time hoodlums to become wealthy and powerful businessmen. Kid Cann and his brothers supplied their own drinking establishments in Minneapolis and also supplied the Chicago Outfit of Johnny Torrio and Al Capone. Speakeasies flourished along Hennepin Avenue and Washington Avenue in Minneapolis.

George Mallon enjoyed a good drink, even a few of them. Prohibition was in effect for thirteen years and George Mallon was never wanting for a drink during the entire period. The fact that he was well liked and well regarded, likely sheltered him from the annoyance of being hauled in for "possession" by local law enforcement. Federal agents tended to focus on the suppliers, which generally connected upstream to organized crime.

The ultimate overseer of Prohibition in Minnesota was S. B. Qvale. He was responsible for District Fourteen of the Department of the Alcohol Tax Unit of the Internal Revenue Service. Qvale and his federal agents officed out of the Landmark Building in downtown St. Paul. His territory included Minnesota, western Wisconsin, North Dakota and South Dakota. Qvale was "the Elliot Ness" of the Upper Midwest and Great Plains.

As mayor of Minneapolis from 1921 to 1929, Mayor George Leach was in an awkward position with respect to Prohibition. There were periodic raids. On April 12, 1922, the chief of the Minneapolis Police Department Vice Department led a raid, which confiscated 130 gallons of moonshine. The officers ceremoniously poured the liquor down the sewer at the corner of Fifth Street and Fourth Avenue South while twenty members of the Women's Christian Temperance Union watched in approval. The *Minneapolis Tribune* included a photo in the morning edition the next day. Of course, everyone knew that had no impact on Minneapolis consumption. On Friday, June 16, 1922, Director Qvale submitted a report to Mayor Leach, which listed 92 soft drink establishments found to be selling intoxicating liquors in a three-month period. Many of those were on Washington Avenue downtown. On December 4, 1922, Qvale unleashed his agents and made arrests at ten different locations on Washington Avenue and on Second Avenue. Through it all, year after year, George Mallon enjoyed his liquor.

There was some talk about another foray into politics by Mallon. In February 1921, the Working People's Nonpartisan Political League endorsed Thomas Van Lear as their candidate for mayor. Van Lear wanted to place his own candidates for alderman in the various wards. The League endorsed George Mallon to run against Alderman Thomas Meagher in Third Ward. Mallon, ultimately, did not accept.

George Mallon still needed to find a way to make a living. The March 18, 1922 issue of the *Minneapolis Morning Tribune* included a small notice back on page 11:

Seeks Bitulithic Member's Job

George H. Mallon yesterday became a candidate for county commissioner to succeed Harry A. Montgomery, one of the "bitulithic trio"... Fourth District... Mallon resides at 703 Sixteenth Avenue South. He was one of Pershing's 100 heroes in the world war.

"Bitulithic trio" referred to Montgomery, A. R. Ferrin and C. B. Wadell to whom the *Tribune* gave considerable press in March as to the three men, who "... awarded the Rockford "luxury" Road paving contract". The *Tribune* dedicated the first two columns of the front page of the March 13 edition, asking "Why?" Minneapolis city alderman asked the governor to investigate the contract, which Montgomery, as chairman of the Hennepin County Board of Commissioners, awarded. In response, the trio organized road rallies to support the expenditures. George Mallon, never one afraid to enter a fight, threw his hat into the ring to right the wrong. Mallon and Montgomery knew each other. They served together in the previous year on the War Records Commission for Hennepin County's contribution to the statewide effort.

Minneapolis Labor Review was quick to praise Mallon while the Montgomery camp attempted to re-position:

> George Mallon has always been known as a man, who meant what he said and was not afraid to say what he thought. It is not a case with him of being friendly to the workers, he is one of the workers with their interests at heart.

Despite his latest political campaign, George Mallon remained active in the local veteran community. He was present at the Elks Club on May 5, 1922 when Theodore Wold, chairman of the Liberty Loan committee, presented the Third Liberty Loan Flag to the Theodor Petersen Post of the American Legion. Wold was a prominent banker, whose son Ernest, an aviator, was killed in action over France. The Theodor Petersen Post, the first Legion post in Minnesota, was named for a sergeant in the medical detail of the 151st Field Artillery, who died of gas exposure on March 5, 1918. Mayor George Leach, who commanded the 151st, received the flag for the post. He told the crowd that the people of Minneapolis originally sent the flag to his battalion in France and that they flew the flag in battle. Leach also urged more zealous care for those who came home disabled.

George Mallon addressed the willingness of the folks back home to support the war effort:

> We soldiers did not win the war. The people, who fought our battles here at home, did more than we did. The men in the front-line trenches don't have the hardest time in war. Give me that experience every time. My wife, my mother, my sister suffered a thousand times more in mental anguish that I ever suffered when I was shot down.
>
> The comrades who couldn't get over, too, were more to be pitied than criticized. And the men, who couldn't have the privilege of fighting on the front, but gave us the financial and moral support at home, they have our lasting gratitude.

The state commander and the former state commander of the Legion also spoke, as did the present and former commanders of the Theodor Petersen Post. When the flag was unfurled to reveal Petersen's name on the flag, Mallon pointed to that and said, "This man gave his life, all he had to give. No man can do better."

George Mallon remained involved with the World War Veterans, which was far less of a threat to the American Legion than they had been just two or three years earlier. The Grand Army of the Republic and the American Legion directed the Memorial Day activities in May 1922. Captain Mallon marched with the World War Veterans in the grand parade.

In June 1922, officials of the Minnesota Historical Society announced its interest in securing for the state the German anti-aircraft gun, which Captain Mallon and his band captured in September 1918. U.S. Congressman Walter H. Newton of Minneapolis was in dialogue with the War Department. John Bowe and the Veterans of Foreign Wars actively supported the effort. The gun was at the Rock Island Arsenal in northern Illinois. According to reports, the Liberty Memorial Association in Kansas City also hoped to secure the gun. Minnesota papers across the state carried the addendum, "… but it is felt that Minnesota should have the prior right." It seemed that the battle of rights to George Mallon continued. Minnesota had no better luck than Missouri. The gun remained at Rock Island Arsenal.

By late summer 1942, the United States was engulfed in war to defend freedom and democracy. The battle against Japanese forces for control of Guadalcanal had just begun and the Americans were preparing to land in North Africa to fight Germany. The outlook was grim with the ultimate outcome of the war nowhere near apparent. The United States scrambled to marshal its resources. Rock Island Arsenal held a vast array of artillery pieces, shells and cannonballs. The government disposed of thirty-six artillery pieces as salvage scrap for the war effort. Among these were several field guns from the Mexican-American War and the American Civil War. The arsenal also sent to scrap the 370mm anti-aircraft gun manufactured by the noted German arms company Friedrich Krupp. George Mallon, even in death, contributed to the war effort in the Second World War. A Rock Island (Illinois) newspaper article dated September 11, 1942 included a photo of a massive long-gun, a ten-ton 155mm German howitzer being scrapped. Perhaps that was also an artifact from Captain Mallon's long gone days of military glory.

As early as June, the people of Minneapolis were picking sides for the November election. *The Minnesota Messenger*, a Minneapolis-based African American newspaper, wrote:

HONOR VETERAN A CANDIDATE

> Capt. Geo. H. Mallon, one of Pershing's 100 Heroes, is a candidate for county commissioner. He will get many of the Negro votes on the north side. Capt. Mallon is on the labor ticket and is a friend of the laboring men and deserves their vote.

Richard B. Montgomery, editor and publisher of another African American Minneapolis-based newspaper, *National Advocate*, regularly endorsed a straight Republican ticket. However, as early as June 10, Montgomery and attorney Harry L. Scott, were actively speaking on behalf of Mallon for the position of county commissioner. The paper again supported the ticket of Preus and Collins for governor and lieutenant governor, as it had done in the 1920 campaign against Mallon.

By the fall 1922, George Mallon's campaign for County Commissioner of Hennepin County was underway. The Mallon campaign held a campaign rally in the new Labor Lyceum at James and Sixth Avenues North on Sunday evening, October 22. As always, Mallon's opponents tried to taint him as the radical leader of the World War Veterans, the radical leader of Nonpartisan League, even as Mallon the Red. The headline of an article in the October 27, 1922 issue of the *Minneapolis Labor Review* read:

Slander Drive Opens Against George Mallon
Orders to "Get" Labor Candidate Given By Interests
Against Mallon

The *Review* wholeheartedly praised and defended Mallon:

No person has ever heard George Mallon boast of his patriotism. Mallon's thousands of friends don't think that they have to say much about it. It is one of those things as established as the pyramids ... It was hardly imagined that Rockford Road Montgomery, still sticky with bithuletic, was one of the auditors who let Hanke get away with a quarter of a million of the taxpayers' money, would attempt to retain his office by swathing himself in the flag and attempting by innuendo to question Mallon's loyalty . . . Montgomery, the individual who was warming a county commissioner chair while Mallon was capturing enemy machine gun nests with no weapons save his bare fists

"Rockford Road" referred to a major highway project, which connected Robbinsdale, home of State Congressman Tom Girling, with Hamel, ten miles to the west. Girling was running for re-election for his fifth term. He was in favor of rural highways and used his influence with county commissioners. He also owned a fleet of buses, which served the outlying towns. Henry Hanke became treasurer of Hennepin County in 1906 and continued in that position until mid 1922. On August 16, 1922, Hanke began serving a sentence of "one to ten years' hard labor" in Stillwater State Prison for embezzlement. County Attorney (later Governor) Floyd B. Olson prosecuted the case. Everyone in county and local state offices did their best to distance themselves from Hanke and their enemies did their best to tie them to Hanke as best they could.

The election took place on November 7, 1922. George Mallon defeated Harry Montgomery and won election as county commissioner. The board, acknowledging Captain Mallon's leadership, promptly elected him chairman. He served in that capacity for four terms, a total of eight years. He was as fearless and determined in politics as he had been as a field commander.

In the same election, Republican Governor J. A. O. Preus

defeated the Farmer-Labor challenger Magnus Johnson. Louis Collins was re-elected Lieutenant Governor. Henrick Shipstead, Mallon's old friend and running-mate from the Nonpartisan League, won the U.S. Senate seat, running on the Farmer-Labor ticket. Shipstead defeated Republican incumbent Frank B. Kellogg, who had been in office since 1917. Kellogg recovered and became U.S. Secretary of State under Presidents Coolidge and Hoover. Shipstead served as U. S. Senator for Minnesota from 1923 through 1946.

Senator Knute Nelson died in office in 1923. The Farmer-Labor Party ran Magnus Johnson for the vacant U.S. Senate seat. Republicans scrambled to somehow retain the seat. Johnson, a former Minnesota state congressman and senator, was popular. The one man, whom Republicans felt could beat Johnson, was Governor Jake Preus. Days before the election, a strange letter was sent out to Minnesota veterans. Among the recipients was George Mallon. *Minneapolis Labor Review* published his caustic reply.

> Your letter received asking me to vote for J. A. O. Preus for senator in order to make Louis Collins governor. You also said this would please ex-servicemen. Let me inform you, comrade, that camouflage is used by real soldiers against the enemy, not against one's friends. Don't use Collins as a foil for others. Although we differ politically, Louis Collins is my friend and I resent this method of using his name for political purposes. I might also add that the election of a United States Senator now is much more important than the advancement of another to the governor's chair. Yours in comradeship, George H. Mallon.

The Farmer-Labor party, from the roots of the Nonpartisan League, controlled both U. S. Senate seats for Minnesota. The two senators became the foundation for the future growth of the Farmer-Labor Party, the predecessor of Minnesota's present-day DFL Party.

George Mallon's county commissioner position paid well and allowed him new-found prestige and position. He and Effie and their two little boys made their home at 1958 Beard Avenue South in Minneapolis. Across the street was the southern edge of Minikahda County Club and just a few blocks to the east was the southern shore of Lake Calhoun, one of Minneapolis' beautiful urban lakes.

More than 10,000 Minnesotans belonged to the American Association for Recognition of the Irish republic. The British agreed on December 6, 1921 to the establishment of The Irish Free State within the year. The *Minneapolis Tribune* announced that the paper would be publishing through a syndication the recollections of Irish leader Michael Collins on twelve consecutive Sundays, beginning January 29. George Mallon and other Irish Americans closely followed the series. That euphoria vanished when the headline of the August 23, 1922 issue of the *Tribune* read, "Michael Collins is Slain." Nonetheless, The Irish Free State came into being on December 6, 1922.

Streetcars efficiently served the Twin Cities in the first two decades of the 20th Century. 1922 was the peak year for streetcars. From that year on, more and more automobiles were on the streets of Minneapolis and St. Paul. In 1923, George Mallon paid $1,436.40 for a 23-45 Buick Touring Car. The price included an extra $26.40 additional, which Mallon paid for bumpers, an accessory at the time. The car was splendid looking vehicle, as well as capable of performance. Wood spoke wheels, a canvas top and leather upholstery were among its features. Buick, having survived the post-world war recession of 1920-1921, began nearly a decade of growth. Buick built 45,000 touring cars of the 23-45 series in 1923. Mallon's six-cylinder touring car was as much fun to be seen in around town as it was to drive.

George H. Mallon and sons Robert & George
Courtesy of Rob Rolfe

Tragedy struck the Mallon family in September 1923. Ollie, George's little brother (of twenty-one year's difference in age), lived in Greybull, Wyoming. He moved out to Wyoming before 1917 and found work with the Chicago Burlington and Quincy Railroad. Ollie became an engineer at 21 years of age in 1919. However, by 1923, as a result of personnel cutbacks, he worked as a fireman. On September 27, 1923, CB&Q Passenger Train #30, on which Ollie was working, was fifteen miles east of Caspar, Wyoming, enroute to Denver. The train raced through the dark of night toward the bridge over Coal Creek near Lockett, Wyoming. A flash flood weakened the bridge, which crossed what was normally a dry creek. As Train #30 began to make its way across the bridge, the structure collapsed and the train plunged into the raging torrent of Cole Creek. The entire crew died, as did more than twenty-five passengers. Some bodies were never recovered.

George Mallon traveled to Caspar to claim his brother's body and brought it back to Ogden for burial in the family plot in St. Patrick's Cemetery. Nearly all Mallons, it seemed, eventually ended up in St. Patrick's Cemetery. That was not the case for Ollie's widow and their only child, four-year-old Bobby. They moved to Los Angeles, California, where Bobby Mallon became one of the Little Rascals in the Our Gang series. He was one of the show's longest lasting characters. They are buried in Los Angeles.

The impact of the world war in Europe continued for years after the Armistice of 1919. Starvation and disease continued to cause deaths. Major General Henry T. Allen served as commander of the American forces during the post-war occupation of Germany. He did not forget what he saw in the period immediately following the war. In December 1923, Major General Henry T. Allen formed the American Committee for Relief of German Children. He arranged for the American Quakers to administer the distribution. The national fundraising goal was $10 million.

Mayor George Leach served as chairman of the Minnesota state committee. George H. Mallon, Chairman of the Board of Hennepin County Commissioners, accepted the position of chairman of the campaign committee for Hennepin County to raise money to support the children of the men whom he faced in battle. The goal for Minnesota was $250,000. Mallon told a reporter:

> I shall do everything I can to help raise all the money possible in Hennepin County. We did not make war on babies and children on the battlefields of France and now that the time has come that Germany realizes she was beaten and needs aid, we should give it to her.

Among those, who joined Mallon on the local committee,

were his friends Colonel W. H. Donahue of the 151st Field Artillery, Z. L. Begin of the VFW and former Minneapolis mayor Thomas Van Lear. James Ford Bell, who would found General Mills in 1928, also joined the committee. Bell was involved with Herbert Hoover's European Hunger Relief Mission four years earlier. Long-time State Senator A. L. Lennon of the Nonpartisan League, who represented Minneapolis from 1915 through 1934 [excluding 1927, which he spent in prison for violating Prohibition], also served on the committee.

The Minnesota committee distributed replicas of mini-life insurance policies re-cast as "Starvation Insurance" with the schedule of 50 cents fed one child for one month and $5.00 fed one child until the next harvest. A brochure, printed in Minneapolis, appealed to the people of Minnesota:

> . . . nothing has ever touched my heart as the condition of the children, blameless as they are for their condition, never having known anything but hunger and want . . . if your child had to eat today only a carrot or a bit of a turnip or some cold potatoes . . . if your baby over two was never to have a drink of milk again . . . if you saw your boy growing up to only half stature, probably with rickets, curvature of the spine or consumption
>
> . . . 5,000,000 German schoolchildren, or 50 per cent of the total number, are undernourished. 2,000,000 of them face starvation . . . The average daily consumption of milk in Munich is one twenty-fifth of a pint per day per capita; in Minneapolis, the figure is one pint . . . America Be Merciful.

The relief effort brought together everyone. George Mallon rode the train south from the Twin Cities to Faribault to speak to the American Legion about the plan to raise funds. H. C. Theopold, chairman of the Rice County Committee, presented

Mallon. The campaign worked closely with the Federal Council of Churches and raised well over $4 million into mid 1924.

Mallon became chairman of the Hennepin County commissioners in 1924. The major construction project of the era in the county began under George Mallon's watch in 1924. This was the construction of the Fort Snelling-Mendota bridge across the Minnesota River Valley. The only way across at this point near the fort was a ferry. The construction lasted from 1924 through 1926. Hennepin County Commissioner Mallon was involved in the project, and many other projects under his purview, throughout the period. The Hennepin County commissioners re-elected George Mallon chairman in 1926.

The magnificent Fort Snelling-Mendota Bridge, when finished, was the longest continuous concrete arch bridge in the world. Thirteen massive concrete arches, each over 300 feet wide, covered a length of 4,100 feet. The state officially opened the bridge on November 8, 1926, dedicating the bridge to the men of the 151st Field Artillery Regiment, who died in the world war. The men of the 151st Association placed a wreath on the memorial bridge. Minneapolis Mayor George Leach, who commanded the 151st during the war, was present. County Commissioner George Mallon knew him well. Governor Christianson untied the golden cord to open the bridge to traffic. Thousands were in attendance, despite the cold weather. The guns of Fort Snelling boomed a salute.

After the dedication on the bridge, three hundred dignitaries attended a dinner at the Nicollet Hotel in Minneapolis. The Nicollet was only two years old and the center of the Minneapolis social scene. The massive twelve-story hotel had 637 rooms. After the invocation, George Mallon, as chairman of the Hennepin County Board of Commissioners, addressed the crowd. He and Walter Wheeler, the project engineer, spoke of the work of financing the bridge and of constructing it. Wheeler introduced C. A. P. Turner, the bridge designer, who also spoke. Mayor Leach, the governor and the lieutenant governor also spoke. Seated at the speakers' table with Mallon was his old friend, Thomas Van Lear. Also seated at the table were the other Hennepin County Commissioners, including W. W. "Pudge" Heffelfinger. Heffelfinger a three-time All-American football player (Yale, 1889-1891), who was a Minneapolis real estate developer.

In 1915, the Minnesota legislature removed the statute banning boxing, which had been in place since 1890. During that twenty-five-year period, boxing events in Minnesota were held only in out-of-the-way venues with small crowds, who learned of upcoming bouts by word-of-mouth. Major public events were held at the Arena in Hudson, Wisconsin, just across the state line. One of the reasons for the restoration was the professional rise of the Gibbon Brothers of St. Paul, Mike and Tommy. Mike was a middleweight and Tommy was a heavyweight. Tommy Gibbons won a newspaper decision in Hudson on February 2, 1915. On July 12, 1915, with boxing restored, Gibbons won a newspaper decision over St. Paul rival Billy Miske at St. Paul Auditorium.

Tommy Gibbons credited Rev. John Dunphy, "J. D.", with helping him launch his professional career in 1911. George Mallon and Father Dunphy knew each other well and it is quite likely that Mallon and Gibbons knew each other. Dunphy even traveled to Montana in 1923 to see Gibbons lose a fifteen-round decision to Joe Dempsey. Tommy Gibbons fought Gene Tunney at the Polo Grounds in New York City on June 5, 1925. It was a St. Paul Irish American facing a New York Irishmen to the delight of Dunphy, Mallon and thousands of other Irish American boxing fans. Tunney knocked out Gibbons in the twelfth round. Gibbons retired from the ring after the fight with a record of 56-4-1. He was thirty-four years old. His brother Mike did not quit boxing until he lost an eye. Tommy Gibbons served as sheriff of Ramsey County 1934 to 1959, winning election for six consecutive four-year terms.

Minneapolis-St. Paul had its share of major boxing events. One of the biggest matches of the era pitted Irishman Bartley

Madden faced Irish American Gene Tunney at the Minneapolis Hockey Arena on September 25, 1925. It was less than three months after Tunney crushed Gibbons' championship hopes. Bartley Madden was a highly-regarded boxer, who was among the contenders to face Jack Dempsey

George Mallon followed boxing for the remainder of his life after leaving the ring.

The June 18, 1926 issue of *Minneapolis Labor Review* announced Mallon's bid for re-election and noted:

> Mallon was candidate for lieutenant governor on the Farmer-Labor ticket in the stormy days immediately following the war and has been the object of severe persecution by some of the bitterest enemies of organized labor during the time he has served as county commissioner.

On the eve of the election, *Minneapolis Labor Review* devoted a column to Mallon as candidate on page 1 of its October 29 edition.

> Incensed at vicious and unfounded attacks that have been hurled at Mallon, the Steamfitters' union at its last meeting unanimously voted to indorse Mallon and issued the following statement:

> For 25 years, we have known George Mallon as an unswerving trade unionist. We have observed him as a consistent trade unionist, on the picket line, as an efficient and honest union executive, courageously answering the call of his country and as a faithful public servant. We watched him return from the war honored, wounded and showered with praise and proffers of high office. We saw him refuse the soft positions offered and instead go out

on the fighting line to face the mobs and hostile throngs for the cause of the organized workers and the organized farmers. Every day that has passed we have come to respect and esteem him more highly. Because we know George H. Mallon so intimately and because of our confidence in him, we urge his re-election to the board of county commissioners . . . Trade Unionists know Mallon was in the thickest of the fight for the workers and the farmers when the battle was most tense.

For all his health problems, George Mallon's first serious incident took place on July 5, 1927. George Mallon had a cerebral hemorrhage, a stroke. He was fifty years old.

Less than two months after Mallon's stroke, Charles Lindbergh flew into the Twin Cities as part of his cross-country flying tour. He successfully completed his historic Atlantic flight and landed in Paris on May 21. He was back in the United States by June 11. Lindbergh began a promotional tour of the United States on July 20. He landed at Minneapolis Wold-Chamberlain Field at 2:03 p.m. on August 23, 1927. Twenty thousand people rushed his plane. He rode in a car with Minneapolis Mayor George Leach and St. Paul Mayor Lawrence Hodgson in what was less of a parade and more of a quick tour. Estimates of the crowd gathered were in the hundreds of thousands and as high as half a million people. Families gathered along the route for hours for the opportunity to see "Lucky Lindy." If the cheering spectators were lucky, they caught a quick blurred glimpse of the famed pilot as the he raced by. The motorcade drove into and passed through Minneapolis then made its way down University Avenue to St. Paul, through St. Paul and to the airport. Lindbergh was at St. Paul airport by 3:30. The *Minneapolis Tribune* wrote, "No parade at all would be preferable." Lindbergh made a quick speech on the merits of aviation and flew off.

The financial world turned upside down when the stock

market crashed on October 29, 1929. Billions of dollars of value, driven up by unrealistic speculation, were lost and the fortunes of anyone in the market were crushed. The collapse of the stock market was not the cause of the Great Depression, but simply a reflection of an underlying economic problem. The nation's economy already was experiencing declines in production. Many Americans already were out of work. In any case, the working class had little disposable income, if any, to invest. They were trying to get by on a day-to-day basis. The Great Depression hit the working class the hardest.

George Mallon posed with H. N. Duff, national commander of the VFW, when he visited Minneapolis in June 1930. *The Minneapolis Journal* carried a photo on the top of page 1 in its Friday, June 30 issue. Mallon did not look good. His poor health was showing, probably a reflection of his stroke.

The position of County Commissioner allowed Mallon exposure to politics and labor. He spoke behind County Attorney (soon-to-be governor) Floyd Olson at the annual picnic of the City and County Employees' Joint Council on Lake Minnetonka in August 1930.

By the fall of 1930, Minneapolis was feeling the effects of what would become known as The Great Depression. In November 1930, George Mallon sought re-election as Hennepin County Commissioner. There was some concern among his friends and supporters. *Minneapolis Labor Review* reported in a page one article on October 31, 1930 that, "Mallon is facing a hard battle for re-election." Mallon's opponent was Arthur Noot, whom the *Review* characterized as, "not a member of organized labor and not indorsed by organized labor." The *Labor Review*'s November 7, 1930 headline read in huge block letters," Farmer-Labor Triumphs in Landslide for Olson." A sub-headline read, "George Mallon Appears to Be Defeated in Fourth District." The Review noted in the article, "This is a matter of regret to the entire labor movement and the Farmer-Labor political movement which Mallon has served with fidelity for many years."

Nearly a decade later, *Minneapolis Labor Review* recalled, "... Noot whose detestible campaign methods even against the late Captain George H. Mallon are remembered by Tenth Ward voters and will never be forgotten by the friends of the famous military hero . . .".

Mallon took his defeat in stride. He placed a sign on his desk, which read, "Lame Duck." That brought a grin to the face of everyone, who went to see him. He extended his thanks to his supporters through *Minneapolis Labor Review*. Nonetheless, the loss of the job, which he had held for the past ten years, just as the Depression began, was not a good harbinger of Captain Mallon's future.

The board elected Pudge Heffelfinger, Mallon's longtime friend, who served as vice-chair under Mallon, as chairman. At Mallon's last county board meeting, Heffelfinger offered a resolution honoring Mallon for his service.

CHAPTER SIXTEEN
HERO OF THE DISTANT PAST

By 1930, the war in Europe was over ten years in the past. Hitler was just beginning his rise to power in Germany. His threat to world peace would not be apparent for three more years. Mallon was still referred to, respectfully, as Captain Mallon by some. The war became a distant memory. Then a flurry of activity again raised his profile.

In 1930, Chevrolet announced a new weekly radio series titled: "The Chevrolet Chronicles." This was part of a national advertising campaign. A full-page advertisement proclaimed, "Hear the personal stories of Americans who have been decorated for valor in action." The emcee of the show was Captain Eddie Rickenbacker, the legendary fighter ace from the world war with 26 kills to his credit. Each radio show lasted thirty minutes. World Broadcasting System produced each episode and distributed it in disc to radio stations across the nation.

Among the featured guests pictured with Captain Mallon were Civil War veteran and Indian fighter Leander Herron, U.S. Navy Lieutenant John McCloy, recipient of the Medal of Honor for the Relief of Peking during the Boxer Rebellion in China and for the Vera Cruz landing in 1944, and Captain H.H. Weimar of world war fame. It was a measure of Mallon's stature at the time that he was included in the series.

The Rickenbacker interview of Captain George Mallon took place on October 28, 1930. The interview began with Rickenbacker saying:

> Like almost every man who was in the service, I am fascinated by war stories, war books and war plays. I have listened to and read about many

amazing feats of daring that were performed by men who were cited for gallantry in action. I have heard about men winning the Medal of Honor by the bayonet and the bomb, by the rifle and the pistol. But, so far as I know, there is only one man who ever won it with bare fists. And he is with us tonight – a big fighting Irishman from Minnesota, who right this moment looks like he would rather be going over the top than getting ready to step up to the microphone beside me. George H. Mallon is his name – formerly captain of Company E, 132nd Infantry – now a peaceful citizen of Minneapolis. Well, Captain, it's zero hour. Are you ready to go over the top?

That brought a laugh from Mallon and the dialogue began.

<u>Mallon</u>: It was in Bois de Forge up in the Argonne. But I'll have to take you back a little to give you a better idea of the action leading up to this particular engagement. Our regiment was a part of the 33rd Division and during the summer of 1918 we had been up with the British. In August, our division was called back into the American army proper to take part in what is known as the Argonne Drive. It was to be our big, final push in ending the war – and it sure was. We came up into the American lines in late August and took up positions at Dead Man's Hill. This hill had been part of the French defenses around Verdun and it got its name from the terrible fighting that had taken place there in 1916. We stayed there until the morning of September 26 when the big drive began all along the line. My company – that is Company E – was to be in the first wave. That night the Allies laid

down a terrific artillery barrage. At five o'clock it lifted and we went over the top. We crossed No Man's Land and rushed the enemy's front lines. We drove them from their positions and carried on toward our objectives, which we reached on scheduled time. And I want you to know, Captain Rickenbacker, that we appreciate the splendid support given us by our comrades in the air on that terrible day. A few days later –

<u>Rickenbacker</u>: Wait a minute, Captain Mallon. You're forgetting something. Your citation says that you captured four 155-millimeter guns, an anti-aircraft gun, eleven machine guns and their crews, which totaled a hundred men.

<u>Mallon</u>: Oh, yes. I forgot. After we passed the first line that morning of September 26th, we kept on toward our objectives. Due to the heavy fog and the woods and the barbed wire entanglements, I had only a small part of my company with me when we got going. After capturing nine machine guns in the woods, we came up to this battery. I didn't have much time to think. Like a flash came the words, which my instructor, John F. Franklin of the Regular Army, used to say to our class at Fort Snelling when we were in training. He said, "When in battle and in doubt as to what to do, do something, push ahead." So, the first thought that came into my mind was the command to charge. I yelled it to the men and then the big job for me was to keep up with those Illinois boys. It was the old Illinois National Guard, you know. Well, using our rifles and bayonets and automatics, we captured those four big guns. Don't ask me about details. It all happened so fast and we were all at it

so hard that it's pretty hard to explain exactly what happened.

Rickenbacker, jokingly, asked Captain Mallon why he used his fist and questioned whether perhaps something was wrong with his automatic. Captain Mallon answered in all seriousness:

> Oh, yes. The Colt was working fine, but the truth is, I wasn't in a position to use it. When we got into the battery, there was one of the enemy behind a cannon and he was shooting at us. I started for him. On the way, I tried to dodge around an enemy officer, but he just grabbed my gun with both his arms and hung on for dear life. Well, there wasn't any Max Schmeling in those days, so I just swung with my left hand and he went down and out for the count. That let me free and we proceeded with the rest of the job. When it was all over we had a lot of artillery and about one hundred prisoners.

Among the most interesting exchanges came about when Rickenbacker asked Captain Mallon how he accomplished all that he did on September 26 without losing a single man. Mallon, always the straight shooter, replied:

> Well, of course, Captain Rickenbacker, you understand at the time that all this happened, it was the government policy either to minimize losses or even not admit them. The truth is, I did lose men. As I remember, I had about twenty men when we started through the woods. Two were sent back with prisoners we picked up along the way. The others were either killed or wounded attacking the machine guns we met while going ahead. So, there were only nine men with me when we captured the battery.

Every time Mallon spoke publicly on the experience, he held the same position:

> I feel that the credit is due to those nine boys more than to me. Naturally, being the one in command and the leader in this particular scrap, I received the lion's share of the credit, as well as the decorations. But I've always felt that the decoration belongs to the men of Company E, 132nd Infantry and is in my care. Whenever I wear it, I always remember those brave boys and feel that they are very close to me. And if any of them are listening tonight, I want them to know that I am still as proud of them as ever and I that I send my affectionate good wishes.

Minneapolis-based Fawcett Publications was the brainchild of Wilford Hamilton "Captain Billy" Fawcett. He founded his enterprise after returning to Minneapolis from the war in France, where he wrote for *Stars & Stripes*. Fawcett produced numerous "pulp" magazines, including the outrageous *Captain Billy's Whiz Bang*, the racy *True Confessions*, *Daring Detective* and *Movie Screen*. These were long before Fawcett's big successes with *Family Circle*, *Mechanix Illustrated* and *Action Comics* (with Captain Marvel). In 1927, Fawcett began a monthly action pulp-fiction magazine called *Battle Stories* in which he re-told the stories of heroes from the world war in fictionalized form with fantastic drawings. The April 1931 issue of *Battle Stories* included "When I Charged Enemy Machine Gun Nests" by Captain George H. Mallon, as told to Rod Russell.

Capt. Eddie Rickenbacker and Capt. George Mallon on the occasion of Mallon's interview for Chevrolet
Courtesy of Hennepin County Library

The publicity continued for Mallon. In November 1931, a syndication, which sold to smaller newspapers across the nation, did a feature for Armistice Day. The title was: "Where are General Pershing's Ten Greatest World War Heroes Today?" The sub header read: "Only Five Celebrate Peace Today." The article included three of the five recipients. One of those was an elderly looking George H. Mallon in a white shirt and tie, standing with his two boys beside him. The piece on Mallon read:

> In the insurance and bonding business at Minneapolis now is George H. Mallon, whose forgotten fame probably will be a surprise to the folks with whom he does business every day.
>
> "I have two boys, 11 and 6 years of age," Mr. Mallon explains as he sits in his office. "And also a wife – I almost forgot her," he chuckles. "No, I don't guess there is much I can say."

The article went on to read: "But War Department records can talk" and followed with Captain Mallon's complete citation.

At about this time, Captain Mallon took part in an event at son Robert's school. Loring Elementary School is located just south of Victory Memorial Drive in Minneapolis. A small group of surviving Civil War veterans and members of the G.A.R. (Grand Army of the Republic) had just dedicated a prominent bronze statue of Abraham Lincoln on Victory Memorial Drive. The Loring students walked to the statue, placed wreaths at its foot and sang patriotic songs. George Mallon, in full uniform and covered with medals, addressed the children, as did N. J. Pelletier of the North Side American Legion Post. A photo appeared in the newspaper with the two veterans and the children with the statue in the background. Standing in the front row of the children was little Robert Mallon, George's second son.

There was still fight left in the old warrior. His spectacles, combed back hair, white shirt and tie and pleated dress pants gave him the appearance of an older man. A family member recalled driving to downtown Minneapolis with George for a shopping trip. Someone pulled into a parking space while George was preparing to back in. She recalled George getting out of his car and walking over to the culprit. Words were exchanged. The next thing she knew, George reached in the car, pulled the man out of his car through the car window (no small feat of strength) and proceeded to throw a series of devastating punches. George Mallon then walked back to the car and drove away, leaving the man lying in the street.

CHAPTER SEVENTEEN
THE FINAL MONTHS

George Mallon was not well in the 1930s. He was only in his fifties, but his life had been a hard one. Some of that was the result of his combat experiences, but not all of it.

Doctors diagnosed Mallon with atherosclerosis, clogging of the arteries, in 1915 at the age of thirty-eight years. In 1921, the doctors told him that he had interstitial nephritis, a disease that would eventually lead to kidney failure. Mallon suffered a stroke, cerebral hemorrhage (uncontrolled bleeding in the brain), in 1927 at the age of fifty years. Heavy drinkers are at much higher risk of a brain hemorrhage than others.

Ironically, Prohibition was in effect for most of George Mallon's life after returning home from war in 1919. Prohibition lasted from 1920 until 1933. In fact, free-flowing alcohol was one of the key elements of the Roaring Twenties and of The Depression. Minneapolis was well known for its many speakeasies during the era. Law enforcement in those days was much more rough-handed on labor organizers and strikers than it was on the proprietors and patrons of illegal drinking establishments.

One of George Mallon's oldest friends in Minneapolis passed away on March 4, 1931. Thomas Van Lear, the Socialist mayor of Minneapolis during the world war, was sixty-two years old. Van Lear was active in the AFL and a close personal friend of Samuel Gompers. Thousands attended the funeral. George H. Mallon was honored to be one of the active pallbearers. They laid Van Lear's remains to rest in Lakewood Cemetery.

Mallon remained active, as best he could, in the Veterans of Foreign Wars and the Disabled American Veterans. The DAV raised money and awareness with the sale of blue Forget-Me-Not flowers in Minneapolis as early as December 17, 1921.

By 1926, they were selling Forget-Me-Nots on September 26, the anniversary of the commencement of the Meuse-Argonne Offensive. It was for his actions on that day that Captain Mallon received his Medal of Honor. At its national convention in 1929, the DAV designated September 26, Argonne Day, and November 11, Armistice Day, as official Forget-Me-Not Days. In the fall of 1931, *Minneapolis Morning Tribune* ran a photo of a young girl pinning a flower on the lapel of a kneeling "Capt. George Mallon" under the caption, "How to Wear Forget-Me-Nots." He did not look well.

One of the major events for world war veterans took place in 1932. The United States was engulfed in The Great Depression. By 1931, desperation was rampant. Food riots occurred across the country, including in Minneapolis and St. Paul. By 1932, tens of thousands of veterans were without work and any means of supporting their families. A movement began for an early payment of a bonus from a bill passed by Congress in 1924. The World War Adjusted Compensation Act provided to world war veterans a certificate redeemable in 1945. The maximum payment was $625 for veterans who served overseas. Veterans argued that they were owed adjusted compensation for loss of pay while in the service of the nation. Detractors referred to this as a bonus. Veterans countered by calling it a "tombstone bonus", since the only way they could receive the bonus before 1945 was to die.

Veterans of Foreign Wars demanded full and immediate payment of the bonus. VFW argued that it made no sense whatsoever to promise a starving man money for food, but not to deliver that for two decades. The national leadership of the American Legion opposed the advance payment of the bonus. Many in the less fortunate rank-and-file of veterans felt otherwise. As a result, the ranks of VFW soared from 1929 on. The Bonus debate highlighted much of what Captain Mallon and the World War Veterans pushed for in 1919 and 1920. That earned them the label of radical.

1932 was the most dismal year of The Great Depression. Veteran Walter W. Waters, a VFW member in Portland, Oregon, was all too aware of the power of the banking and commerce lobbies in the nation's capital. In March, Waters declared a march on Washington, D.C. to press their case. Three hundred veterans from Portland joined him. Their journey ignited a grassroots effort as veterans from across the nation made their way to the nation's capital. It was a movement of hope in a very dark time for millions of Americans. They were not bums. They were, by design, an inconvenience. It was the best way they know to lobby. They represented the vast unemployed of America – veterans, as well as others.

Some argued that the VFW's position set in motion a national march on Washington. The VFW did not officially sanction the march. While both the national American Legion and the national VFW withheld official support for the march, many members of both veteran organizations participated and supported.

The first of the Bonus Army arrived in Washington on May 25, 1932. As many as forty thousand men followed. Walter W. Waters later compiled state summaries from the rosters in his possession of the men of the BEF. He emphasized, "These sheets contain perhaps half of the total who registered . . .". Waters identified 399 veterans from Iowa, 394 veterans from Wisconsin and 311 from Minnesota. The Bonus Army veterans built shacks from scrap lumber, scrap tin and cardboard. They carefully laid out streets, set up kitchens, hospitals and latrines. There was no segregation. They established a library and a post office. They organized classes and entertainment. The Blues was the soundtrack of the Bonus Army. The veterans organized around military discipline. Liquor and panhandling were not allowed. Poorly dressed men walked proudly wearing the medals, which they had earned in the military service.

By July 28, President Hoover's patience was exhausted. Attorney General William D. Mitchell, a St. Paul attorney, ordered the camps cleared. The veterans resisted the Washington,

D.C. police, who killed two veterans and wounded many more. President Hoover sent in the U.S. Army, commanded by Chief of Staff General Douglas MacArthur. They drove out families and burned the shanty towns. *The Minneapolis Labor Review* cited an editorial by Scripps-Howard newspapers, which actually opposed payment of the bonus, but saw wrong in the attack:

> Revolutions cannot be threatened by unarmed men. The bonus marchers were unarmed ... Now is the test of this democracy. If the Government cannot lead peacefully, democracy is gone. If the Army must be called out to make war on unarmed citizens, this is no longer America.

Minnesotan veterans made their way home from the Washington, D.C. debacle. In the fall of 1932, about forty of them camped at 53rd Avenue North and Lyndale along the Mississippi River. On November 5 in the final days of the 1932 presidential campaign, President Herbert Hoover spoke in St. Paul and declared, "Thank God we have a government in Washington that knows how to deal with a mob." That remark played no small role in Hoover losing his re-election bid. Democratic candidate Franklin Delano Roosevelt won the election easily and became president. FDR took office on March 4, 1933 and began to work tirelessly to raise the United States and all Americans out of the depths of the Great Depression.

George Mallon did his best to remain active in labor circles. Minneapolis, despite all Mallon's efforts and that of many others, remained an open shop town. Citizens Alliance, well-funded by major businesses and assisted by law enforcement, continued to meet and crush every union effort with brutal resistance.

The Roosevelt administration delivered some relief to organized labor in June 1933 when FDR signed into law the National Industrial Recovery Act. The legislation gave employees the right to organize and bargain collectively with employers, free from coercion. Workers swelled the ranks of unions and strikes soon commenced across the nation. The garment industry had a long history of poor working conditions and low wages, maintained by tough guys breaking the heads of organizers. On July 20, 1933, the Amalgamated Clothing Workers at Robitshek-Schneider Company in Minneapolis went on strike. That was only the beginning.

In October, Upholsterers Union Local 61 organized and demanded a closed-shop contract with seven Minneapolis furniture companies, all seven of which were contributing members of Citizens Alliance. The union went on strike on October 18. On December 27, to the surprise of the unions, the Labor Board sided with Citizens Alliance ordered an end to the strike. Minneapolis labor turned out in force on the following day to protest. The day began with a meeting at the East Side Eagles Hall, which was packed beyond capacity with thousands of others waiting outside in the streets. *Minneapolis Labor Review* reported that "Captain George H. Mallon" was among the speakers, but did not print any of his remarks. The front-page headline for the *Minneapolis Labor Review* on the following day read:

> Thousands Join in Demonstrations to Aid Furniture Workers
> Great Throng Makes Plain That Masses of Workers are Solidly Behind

> After the meeting the demonstration proceeded to march past the Levin plant, then up to the Grau-Curtis plant and past the Northwestern and McLeod-Smiths and over to the plant of the Brooks Parlor Furniture Company. Here a foolish police order that the marchers were not to be permitted to circle the Brooks plant caused a clash with police.

In the midst of pushing and scuffling, the police gassed the crowd. Among those gassed was George Mallon's friend, U.S. Congressman Ernest Lundeen. The police arrested two unruly demonstrators and the situation threatened to escalate even further. Cooler heads did their best to calm the crowd, while they pleaded with the police. The police, greatly outnumbered and not willing to begin gunning down citizens, released the prisoners. It was, perhaps, Captain Mallon's final combat experience.

The VFW held its annual encampment in late August 1933 in Milwaukee. Among the speakers were Marine Corps Major General and two-time Medal of Honor recipient Smedley D. Butler and outspoken populist Senator Huey P. Long of Louisiana. Butler was unabashed about attacking Big Business and its control of government. Butler was an intensified version of George Mallon and Long was the enhanced form of Arthur Townley. VFW was a far more powerful national vehicle than World War Veterans or Nonpartisan League. George Mallon remained active in VFW, but his health limited his ability to travel and appear.

The VFW elected James Van Zandt as national commander for the coming year. In December 1933, Van Zandt and Butler went on a whirlwind speaking tour – ten cities in eleven days. Among those cities was St. Paul. George Mallon likely met Van Zandt and Butler, but there is no record of such a meeting.

A new labor group gained power in Minneapolis in 1934. The Dunne Brothers, Grant, Ray and Mick, and a core of comrades took over Minneapolis Teamsters Local 574 through a grassroots campaign. Conservatives wasted their time trying to label George Mallon a radical in the 1920s. He was a progressive. The Dunne Brothers, however, were true radicals. They were

Trotskyites, committed to revolution. In the frigid Minneapolis winter of early 1934, Local 574 organized a coal truckers strike. Their timing was excellent and their aggressive tactics shut down the much-needed coal. *Minneapolis Labor Review* of February 9, 1934 noted, "Workers of Entire City Inspired by Militant Effectiveness of Drivers Strike." Coal employers threw in the towel and labor rejoiced. The win swelled the ranks of Local 574. Few of the members were Trotskyites, but truckers saw results and responded by joining the Teamster local.

In early February of 1934, while the coal strike was underway, George Mallon suffered a second stroke. He entered the Veterans Hospital at Fort Snelling. His long history of poor health caught up with him. He retained hope that the Farmer-Labor Party might endorse him once again for the position of Hennepin County Commissioner, but he did not get out of the hospital. On March 1, doctors transferred Captain Mallon to the large Veterans Hospital in St. Cloud, eighty miles to the northwest. The Farmer-Labor Party held its county convention at the Labor Lyceum on Sunday afternoon, March 4. The party chose, instead, Representative Emil Youngdahl. *Minneapolis Labor Review* of March 9 noted, "Captain George H. Mallon showed he had hosts of friends in the strong contest he made for the indorsement against Youngdale."

George Mallon was at the St. Cloud VA on June 15 for his 57th birthday. He apparently could get out of the hospital bed and move around. In late July, George Mallon suffered an attack of epilepsy.

Nobel Prize-winning author Sinclair Lewis appeared at the St. Cloud VA on Saturday morning, July 28. He was born and raised in nearby Sauk Centre and still had family there. The VA director took Lewis on a tour of the hospital. While Sinclair Lewis reportedly visited with some of the patients at the VA, there is no record of who those were. Perhaps George Mallon met Sinclair Lewis.

Sinclair Lewis and George Mallon shared a past with the

Nonpartisan League fifteen years earlier. Lewis and his wife lived in St. Paul and Minneapolis in 1917 and 1918 and in Mankato in 1919 while he researched what would become *Main Street*, the 1920 best seller that established Sinclair's reputation as a writer. *Nonpartisan leaders* frequented their home and the couple attended Nonpartisan League picnics and rallies. Thomas Van Lear likely would have introduced Lewis and Mallon. The novel *Main Street* tells the story of a woman, who moves to a small town and witnesses the rise of the farmers' insurgency that is the Nonpartisan League and the fierce resistance the league faces. Lewis' 1922 novel *Babbitt* subsequently satirized the unthinking conformity of the middle class. In 1934, Lewis had just finished his cautionary novel, *It Can't Happen Here*, about the rise of a dictatorship in the United States. It is difficult to imagine that the hospital staff did not direct Lewis to George Mallon.

While George fought his last fight, a war was underway in Minneapolis. The Minneapolis General Strike began on May 16, the result of a trucking strike called by the local Teamsters. Violence was a daily occurrence in the Market District, even to the point of open battles between hundreds of combatants. Perhaps the presence of Mallon between the two might have calmed the situation. However, his time was in the past. On July 20, known as "Bloody Friday", police opened fire on a crowd in the streets of Minneapolis, killing two and wounding 67. Most of the wounded were shot in the back while trying to flee.

On July 31, George Mallon was found to have peritonitis, an inflammation possibly resulting from a rupture of the abdomen. The doctors at the VA contacted Effie and told her to come to St. Cloud. She brought the boys, George Jr. and Robert, ages 14 and 9, respectively. The three of them were at George's bedside when he passed away on August 2, 1934. The immediate cause of death was hypostatic pneumonia, a collection of fluid in the lungs that often occurs when someone is bedridden. The clinical director noted that contributory causes of death were cirrhosis of the liver, due to alcoholism and general paresis.

George Mallon's grandson, also of the same name, recalled that his parents and the other elders of the family always talked like it was a war wound that ultimately killed the Medal of Honor recipient. Perhaps they were right. We now know that many soldiers come home from war with invisible wounds in their head and in their heart and in their soul. No one can see the wounds, but the family experiences the impact of the emotional trauma of war.

Minneapolis Labor Review published a beautiful tribute to the late Captain Mallon titled "A Brave Man Dies."

> Too seldom is there found in one man both physical and moral courage. Too often it happens that the man, who never flinches physically, may cower when it comes to a test of moral courage. And he, who is unafraid morally, shrinks from physical harm. Captain George H. Mallon, who died last Thursday, combined these two qualities as they may have seldom been combined in any one individual . . . Captain Mallon was also a trade unionist of both moral and physical courage. He could present whatever quality was necessary in the long struggle of labor for justice.

The *Labor Review* could not help but use Mallon's spirit and memory to frame the chaos in the streets of Minneapolis at the time between the truckers local union and the corporate interests of Citizens Alliance.

> On the political field when all Farmer-Laborites were being branded as radicals, reds and undesirables, by the same outfit of highbinders and phony stock sharks who are now assailing the General Drivers

and mobs were raised up to destroy them. Captain Mallon met those mobs unflinchingly and turned them back. He despised and detested the Citizens Alliance. This band of oppressors had attacked him mercilessly and anyone who believes for a minute that any Citizens Alliance attack on a man because they claim he is red or radical is sincere is sadly misled. They attacked and assailed and vilified Captain Mallon with the same bitterness they are attacking organized labor today . . . he never forgot that the greatest conflict was not between nations, but that being made by the exploited and oppressed for freedom and peace and plenty.

The violence in Minneapolis ceased on August 22. In the end, the unions overcame Citizens Alliance and ended the open shop policies of Minneapolis. George Mallon would have been pleased.

The funeral service for George Mallon was held at Ste. Anne's Catholic Church, just off West Broadway in North Minneapolis. Rev. John Dunphy, pastor of Ascension Church, was the celebrant of the funeral Mass, along with his two assistants from Ascension, Rev. Thomas Meagher and Rev. Francis Nolan, and Rev. James Coleman, pastor of St. Francis Xavier Church, Rev. Max Matz, pastor of All Saints' Catholic Church, and Rev. Thomas O'Brien. The Irish Americans were well represented.

Father John Dunphy was born in County Kilkenny, Ireland. He came to the Twin Cities in response to a call by Archbishop John Ireland for Irish priests for the archdiocese. From 1900 to 1921, Dunphy was Vice-Rector, Prefect of Discipline, Dean of Studies, and Athletic Director at St. Thomas College in St. Paul. He instilled respect for authority and discipline in students and fostered athleticism and competition. Former students referred to him as "The King". Father Dunphy shared George Mallon's passion for boxing. Renowned heavyweight fighter Tommy

Gibbons of St. Paul credited "J. D." with helping launch his career. Rev. Dunphy traveled to Montana in 1923 to see Gibbons lose a fifteen-round decision to Joe Dempsey. By 1934, Gibbons was in the insurance business, like George Mallon.

George Mallon's large circle of friends filled the large church to overflow with as many as one thousand outside. Thousands joined the long funeral cortege that made its way to St. Mary's Cemetery. The cemetery opened in 1873 as the parish cemetery for the Church of the Immaculate Conception, which became the Basilica of St. Mary.

Among Mallon's active pallbearers was Ben Neff. Neff was a Russian Jew, who went absent without leave from the Imperial Russian Army and made his way to Canada and, finally, to North Dakota. The horrific pogroms of 1905 were still a recent memory for the Jews of Russia. Private Neff served with distinction in the world war with the 1st Infantry Division and received the Distinguished Service Cross for valor near Exermont, France on October 4, 1918. He was gassed not long after that fight and spent several weeks in the hospital. Ben Neff moved to Minneapolis after the war and came to know Mallon well through their memberships in both Minnesota Disabled American Veterans and Veterans of Foreign Wars. Neff was commander of Minnesota DAV and vice commander of national DAV.

Another active pallbearer was Colonel William H. Donahue, a Minneapolis attorney. He served with the 151st Field Artillery Battalion under George Leach. Donahue was a recipient of the Distinguished Service Cross. On March 5, 1918, then Lt. Colonel Donahue rushed into the quarry in which Battery C was positioned. The Germans were zeroed in on the quarry and wreaking havoc on the men of Battery C. Donahue disregarded his personal safety and assisted the officers. Minneapolis newspaper *The Irish Standard* gave Donahue considerable coverage during and after the war.

Herbert Campbell of the Disabled American Veterans was an active pallbearer. The last three of Captain Mallon's active pallbearers were members of the Veterans of Foreign Wars, all union men: Herb Milbrath, Fred Betzold and Robert Rubinger. Betzold was an executive with the Plasterers Union. Rubinger was an Executive with the Plumbers Union. Rubinger served in the war with the 36th Division in the world war and saw considerable action.

Honorary pallbearers included Bert Baston. Lieutenant Baston of the 5th Marine Regiment received the Navy Cross for his heroism near Belleau Wood on June 6, 1918. He owned a Chevrolet dealership on Hennepin Avenue in Minneapolis. An All-American football player for University of Minnesota before the war, Baston was an assistant football coach for the U at the time of Mallon's death.

State Senator Robert G. Marshall was an honorary pallbearer. Lieutenant (later captain) Marshall of the 58th Infantry Regiment received the Distinguished Service Cross for heroism near Bois-du-Fays, France, 4 October 1918. He was a well- regarded man in town, who held the line and survived horrific shelling during the war.

Kimon Karelis was an honorary pallbearer. He entered the military service as a young Greek immigrant and came home from France with the Distinguished Service Cross. Private Karelis served with the 15th Machine Gun Battalion of the 5th Division. On September 12-13, 1918, Karelis continued fighting after being seriously wounded.

Ted Wallin, who served in the world war in a medical detachment, and other veterans were also honorary pallbearers. Dr. R. R. Hein, past state commander of the Minnesota VFW, was an honorary pallbearer, as was VFW member John Sokolowski.

Governor Floyd B. Olson was among the honorary pallbearers at Mallon's funeral. The Farmer-Labor Party acknowledged its debt to the Nonpartisan League. The Farmer-Labor Party rose to prominence in the final years of George Mallon's life. He was alive and well when Floyd B. Olson became governor of Minnesota on January 6, 1931. Long-time Minnesota

State Senator A. L. Lennon of the Nonpartisan League was another honorary pallbearer, as was Brigadier General Ellard Walsh, Adjutant General of Minnesota. Brigadier General David Stone, commandant of Fort Snelling, was an honorary pallbearer. Stone, like Mallon, was a veteran of the Philippines War.

Other honorary pallbearers included Judge Luther Youngdahl, who became governor in 1947, and District Court Judge Levi M. Hall. Hall attended Officers Training at Fort Snelling with Mallon in 1917. Originally assigned to the 33rd Division with Mallon, he later transferred to the 90th Aero Squadron, which flew tactical recon over 33rd Division in the Meuse-Argonne offensive. Hall was a very active member of the American Legion, as well as the Veterans of Foreign Wars.

W. W. "Pudge" Heffelfinger was a three-time All-American football player for Yale for 1889-1891. He returned to Minneapolis and, beginning in 1924, served on the Hennepin County Board of Commissioners. He served with George Mallon on the Board and well beyond the end of Mallon's time on the Board.

The pallbearers and many friends formed a line at the cemetery gate and marched to the grave, accompanied by a military band playing a funeral march. The Russell Gaylord Post of the Veterans of Foreign Wars and the Disabled American Veterans, to which Mallon belonged, presided at the grave.

The Russell Gaylord VFW Post to which George Mallon belonged was named for a Minneapolis resident, a graduate of North High School and University of Minnesota. Gaylord was nearing the end of his second year at Harvard Law School when war broke out and he enlisted in the army. He became a lieutenant in the 18th Infantry Regiment, 1st Division. Lt. Gaylord was killed in action at Villers-Tournelle on April 28, 1918. His remains lie in the Somme American Cemetery in France. Mrs. Gaylord, the wife of Russell's brother, was the soloist at the grave. A squad from the Third Infantry Regiment at Fort Snelling fired volleys over the grave and a trumpet sounded taps.

The *Minneapolis Labor Review* reported on the Mallon funeral. The headline read:

GREAT THRONG AT FUNERAL OF CAPT. MALLON

Daily Papers Failed to Mention His Activities on Behalf of Organized Labor. Carry No Description of Funeral Services. Declared He Prized His Union Card As Much As Any Decoration Received.

Veteran of Foreign Wars' magazine **Foreign Service** reported on the deaths of its members each month in a column titled "Taps." Mallon's passing merited an article titled, "Death of a Hero" in its September 1934 issue. The article listed Captain Mallon's many medals and honorifics and included the citation for his Medal of Honor.

> ... the VFW lost an outstanding member ... He was particularly active in veteran affairs and much in demand as an official of parade committees and for various civic and patriotic events ... Post No. 159 was in charge of burial services which included the full VFW ritual.

On the subsequent Armistice Day, three months after Captain Mallon's passing, the Veterans of Foreign Wars formally installed the George H. Mallon Post No. 3144. The dedication booklet featured a photo of Captain Mallon with a full-page biography. Charter members were:

F. M. Babb	Edmund H. Harrigan
Frederick Baillargeon	Carl Hedblad
Arnold Barrette	Harry A. Johnston
Gust Benson	Arthur O. La Palme
Arthur G. Berg	Ernest Lundberg
Arthur Bethke	Sigurd Ness

Paul Bodthe	George Ramsey
Alfred Bolin	Alfred Rath
Anthony Butcivitch	Henry Reil
Oscar Catlin	Robert Rubinger
Julius H. Cavilla	Owen Ryan
James Ray Cooper	Bryan Silgjord
Joseph Cronin	Guy E. Thornton
Rudolph Ettle	Raymond Welch
Carl Frederickson	Erick G. Westerberg
Daniel Gerdes	Wilfred F. Plaisance

Robert Rubinger had been a pallbearer at Mallon's funeral. He also was one of the organizers of the Mallon post. On behalf of the Steamfitters Union, he presented the new post with a silk American flag. Unfortunately, Rubinger's remarks on behalf of his friend, George Mallon, were not saved for history's sake.

The Mallon Post was at 2230 Washington Avenue North, north of West Broadway, for many years. An event announcement in 1937 stated, "Members of organized labor will find a distinctly union atmosphere at this card party and dance as the members of the Captain Mallon post are all members of organized labor." In 1951, the Post leadership challenged a move by Minneapolis City Council to amend the charter and reduce the number of aldermen. The VFW suggested that any such motion be postponed until all those mobilized for the Korean War returned home. The story was the main headline on page 1 with a second headline, which read, "Post Follows Policies of Brave Leader." The article ended with reference to "Fearless Captain George H. Mallon" and his brilliant record as a labor man.

The Minneapolis Trucking Strike of 1934 ended on August 21, 1934 after Citizens Alliance agreed to a mediated settlement, which incorporated the Teamsters' major demands. It was an unprecedented victory for labor, not only in Minneapolis, but across the nation. Everyone followed the events and everyone understood the implications for unions. The outcome would not have been possible without The Great Depression and the heartless Republican administration of Herbert Hoover followed by Franklin Delano Roosevelt and the early elements of his New Deal. The National Industrial Recovery Act in June 1933 guaranteed the right of workers to form unions and provided for collective bargaining. The innate ability of Trotskyists to organize the Teamsters and to execute a strike set up the landmark Minneapolis Trucking Strike of 1934. In the end, it was the Teamsters willingness to use violence in response to violence, which defeated Citizens Alliance and opened Minneapolis to labor unions.

There is little doubt that the determined efforts of so many labor men in the 1920s before the Trotskyists were critical to setting the stage for 1934. Among the foremost was George H. Mallon. His final days were during this great fight and he did not live to see the outcome. Those, who once thought Mallon to be radical, did not anticipate the Trotskyists.

CHAPTER EIGHTEEN
FORT SNELLING NATIONAL CEMETERY

The story of the final resting place of the earthly remains of George Mallon begins with the story of an amazing woman by the name of Theresa Ericksen.

Ericksen, a Norwegian immigrant as a child, was an 1894 graduate of the School of Nursing at Minneapolis' Northwestern Hospital. In 1898 during the Spanish American War, the Minnesota Daughters of the American Revolution sent Theresa and others to serve for three months at the U.S. Army Hospital at Chickamauga, Georgia. The conditions were deplorable. Miss Ericksen distinguished herself caring for the soldiers of the 12th and 14th Minnesota Infantry, who were stricken with typhoid fever. Theresa accompanied several recovering patients back to the Twin Cities. Dr. Parks Ritchie, dean of the University of Minnesota Medical Department, summoned Ericksen to his office in early October 1898. The dean's son, Captain Harry Ritchie, an assistant surgeon with the 13th Minnesota Volunteer Infantry, requested that his father find some nurses for the Philippines. Theresa Ericksen was the first Minnesotan woman to be chosen. Some accounts assign her the honor of being the first Army nurse in the Philippines. The Surgeon General approved and she arrived in Manila in February 1899. Theresa endeared herself to many soldiers in her tireless service. After serving at a hospital in Manila, she was sent to a field hospital at Dagupan in northern Luzon. There was a constant threat of guerillas and all nurses trained in firing pistols. One night, a soldier, reeling from fever, was threatening to jump from the third floor of the hospital. Nurse Ericksen managed to save the patient and, small as she was, began to carry him to his bed. The rotten floor gave way under the force of their combined weight and she fell through, shattering her kneecap. She walked with a cane for the rest of her life. Theresa remained in the Philippines until August 1901. She returned to the Philippines in 1904 to work for a time at Manila's new hospital.

Theresa Ericksen applied for the Army Nurse Corps in April 1918. The Surgeon General's Office rejected her due to her age. She was fifty-one years old. The Red Cross recognized her experience and offered Ericksen a position as Reconstruction Aide Masseuse in June. She accepted, sailed to France and soon set to work as a nurse. There simply were not enough nurses to meet the demand. Ericksen first served in an Army Hospital, which received wounded from the fighting at Chateau-Thierry. She cared for soldiers and Marines, who had lost part of their faces. That may have been American Red Cross Hospital # 111, an emergency evacuation hospital in the Hotel Dieu in Chateau Thierry.

After three months, Ericksen transferred to Hospital Violet near Lyons to work in a hospital for war orphans with contagious diseases. In October 1918, deadly influenza raced through the nearby officers' training camp at La Valbonne. Hospital Violet evacuated the children and filled with sick and dying doughboys.

Theresa Ericksen returned to Minnesota a war veteran in 1919 and took a position as a school nurse in Hubbard County. She was one of the founding members (and the only woman) of American Legion Post #212 in Park Rapids in September 1919. In the early 1920s, Ericksen joined A. R. Patterson Post #7 of the Veterans of Foreign Wars in Minneapolis. She reportedly was the first women to become a member of VFW. Even as Miss Ericksen embarked on a career as a public health nurse, she was well known for her active role in veteran affairs.

A newspaper report from the 37th Annual National Encampment of the United Spanish War Veterans in San Antonio in 1935 read: "One of the most beloved figures attending the Spanish American war encampment is Miss Theresa Ericksen, "the little nurse from Minnesota," beloved of thousands of veterans of the Spanish-American War, the Philippine Insurrection. It was this humble nurse, who sought an appropriate resting place in Minnesota. The chain of events, which she set off, began just months after the funeral and burial of Captain George H. Mallon in St. Mary's Cemetery in Minneapolis.

F. W. Pederson, Commandant of the Minnesota Soldiers' Home, wrote U.S. Congressman Ernest Lundeen on February 2, 1935 on behalf of Theresa Ericksen with a request that she be allowed to be buried in the post cemetery at Fort Snelling.

Fort Snelling had been a presence in the Upper Midwest since 1820. The Post Cemetery was the final resting place for 680 soldiers and family members, most of whom died while assigned to the post. This included veterans of the Civil War and the Indian Wars. Lundeen passed on the request to the Secretary of War of the United States, who responded on April 8 that only those who were on active duty or on the Retired List and members of their immediate families were entitled to burial in post cemeteries.

Theresa Ericksen was no stranger to bureaucracy. When she became seriously ill in 1920 and applied for War Risk Compensation, the government ruled that "because she served under Red Cross and not in the Army Nurse Corps, she is not entitled to any compensation or state bonus and not even to the sixty dollars bonus given by the government to the nurses at the time of discharge." At the beginning of 1935, Miss Ericksen was sixty-seven years old. However, the government ruled that she was not eligible for any pension connected with her world war service, because the Surgeon General's Office rejected her from acceptance into the Army Nurse Corps.

F. W. Pederson, commandant of the Minnesota Soldiers Home, responded in earnest on April 30, 1935 to Lundeen:

> I am just in receipt of your letter of April 12th regarding the request of Theresa Ericksen, our little Minnesota Nurse, who wanted to secure her last resting place in the Post Cemetery at Fort

Snelling, also your copy of a letter from the War Department. We note that they rule against her request, for which we are sorry. I have written her and given her the information. We also note you enclose a list of eighty-six different National Cemeteries in which she might find a resting place. Of course, you understand she would not want to be buried in any one of these as she is so thoroughly a Minnesota Veteran that she will want to be buried in Minnesota and we will find place for her. She has remarked that she could be buried in Arlington as nurses have a separate plot there, but that would be a long way from Minnesota and you know we all have a certain sentiment. In the meantime, this brings on another thought and that is why not have a National Cemetery in the State of Minnesota.

This letter became the catalyst for a serious and long overdue discussion as to the establishment of a national cemetery in Minnesota for the Upper Midwest.

The January 1940 issue of *All Veterans News* stated, "Theresa Ericksen of the Minnesota Soldiers' Home, one of the first workers in behalf of the Fort Snelling National Cemetery, is now with tireless efforts keeping a continual flow of letters to our Congressmen and Senators to assist in the development of our Cemetery."

Dr. John E. Soper, a medical officer with the 151st Field Artillery in the world war, was stepping down from his position as President of the Minneapolis Memorial Day Association in 1935. He organized a group, which he called the National Cemetery Committee, in that same year of the Ericksen issue. Soper expended considerable effort, including numerous trips to Washington, D.C., to get the national cemetery approved. The many veteran organizations in Minnesota lent considerable support in the lobbying effort. There was reference to the

Re-interment of George Mallon
in Fort Snelling National Cemetery
Courtesy of George Mallon

requested cemetery as "a second Arlington".

The late George Mallon's friend Ben Neff, recipient of the Distinguished Service Cross, played an active role in the effort to have a national cemetery established in the Twin Cities. It is not known for certain if Captain Mallon ever discussed with Neff the need for a national cemetery in Minnesota, but it seems possible. Other members of the committee included John Seaberg, who saw action in the Meuse-Argonne as a corporal with the 604th Engineers in France.

On June 23, 1936, Congress approved legislation, which established Fort Snelling National Cemetery. The legislation was amended in the following year, authorizing the War Department to allocate 180 acres from the vast Fort Snelling Military Reservation for the new national cemetery (additional acreage was added in 1960 and, again, in 1961). The federal government's WPA (Works Progress Administration) program provided initial funding and labor. The State of Minnesota later contributed some money. By March 28, 1938, as the project approached completion, Minnesotans honored Dr. Soper at a banquet at the Nicollet Hotel in Minneapolis.

On July 5, 1939, nine days before the formal dedication of Fort Snelling National Cemetery, local veterans reinterred Mallon's remains as the first burial in Minnesota's own hallowed ground. The military rites at the grave were under the auspices of both the Russell Gaylord VFW Post and the George Mallon VFW Post. J. A. Stern of the Gaylord Post shared some words:

> Like a brave man he marched away, like a strong man, he returned to the duties of civil life. The red of our country's flag was made still redder by his heroism.

Chaplain J. A. Greenwood of the Gaylord Post invoked a blessing. George W. Anderson, junior vice commander of Minnesota VFW, placed a sprig of evergreen upon the casket and William Plaisance, senior vice commander, placed a spray of white flowers. A. R. Ulstrom, officer of the day, placed a laurel wreath on the casket. A soloist sang the VFW memorial song *Sleep, Soldier Boy, Sleep* (written by Dorothy Alexander in 1926). The rifle squad fired volleys and the casket was lowered into the ground as a bugler played Taps.

Captain George Mallon's headstone is in Section DS (for Distinguished Service) in gravesite 1-S. The main road of the cemetery, leading from the main flagpole, runs through Section DS and is designated Mallon Avenue. Not far from the captain's stone is that of his oldest son, Robert Curry Mallon, a World War Two veteran, who died in 2009. His remains lie in Section B, gravesite 139-2.

Perhaps an appropriate epitaph would have been as *Minneapolis Labor Review* wrote:

> He hated war with its carnage and its heroes and its profits for those who did not fight. He hated war even though he did not fear it.

Photograph by the author

Emil Holmes, a Minneapolis veteran of the world war and a national official of Disabled American Veterans, addressed a gathering of the Central Labor Union in May 1940. By that time, all Americans could sense the pending U.S. involvement in the war in Europe and Asia. *Minneapolis Labor Review* reported in its May 24, 1940 issue that Holmes "… paid an eloquent tribute to the late Captain George H. Mallon." The old warrior's presence in the new national cemetery helped to solidify memory of him.

Three months later, on August 15, 1940, U.S. Senator Ernest Lundeen of Minnesota addressed the United States Congress on the history of the Farmer-Labor Party with which he was affiliated. He mentioned Mallon for his role in sustaining the Farmer-Labor Party at the time:

After the primaries, King withdrew in behalf of Shipstead, who, with Mallon and Sullivan, then filed by petition for the same offices they sought in the Republican primary, but this time as independents with Farmer-Labor Labor endorsement. These rather clumsy arrangements were the best that could be done while the Progressives were still undecided whether or not to keep the Farmer-Labor Party alive.

George H. Mallon is known to some historians in the Twin Cities. However, as with most heroes of the First World War, he is largely forgotten. Those, who do know of Mallon, primarily know him as the Medal of Honor recipient that he was. George Mallon was so much more than that.

Among those buried in Fort Snelling with Captain Mallon are:

U.S. Congressman Ernest Lundeen became U.S. Senator Lundeen. He died in a plane crash on August 31, 1940 and was buried in Fort Snelling National Cemetery in Section B, gravesite 140-S. Theresa Ericksen died on September 1, 1943 at the Veterans Hospital after a long illness. She was buried in Fort Snelling National Cemetery on September 2, 1943. Veterans of the 13th Minnesota Volunteer Infantry fired a salute at graveside. Her remains lie in Section A-11, gravesite

1884. Her obituary identified the little nurse as being "generally credited with having initiated the idea of the Fort Snelling cemetery and worked indefatigably for its establishment."

Ben Neff remained active in Minneapolis with the Disabled American Veterans as late as 1954. He died April 23, 1962 at the Phoenix VA Hospital. His obituary in the St. Paul *Dispatch* was brief and mentioned no family. His wishes were to be returned to the Twin Cities and buried in Fort Snelling National Cemetery. His remains are buried in the DS (Distinguished Service) section of the cemetery, not far from his friend George Mallon, in gravesite 11N. John W. Seaberg, who served in the cemetery committee with Ben Neff, died in 1970 and is buried in Section L in gravesite 1163.

Dr. John E. Soper, captain with the 151st Field Artillery in World War One, died March 24, 1956. He is buried in Section B, gravesite 137-1 in Fort Snelling National Cemetery.

Among those Americans, who fought alongside Captain Mallon at Hamel was Private Fred R. Wilkins of Company A, 132nd Infantry Regiment. He received the Distinguished Service Cross for attacking a machine gun on his own, killing and driving off the crew with hand grenades and capturing the gun. He is buried in the Distinguished Service (DS) section of Fort Snelling National Cemetery in gravesite 77-N.

Among those, who served under Captain Mallon in Company E in 1918 was Private Andrew Ricker of Minneapolis. He died in 1970 and is buried in Section M, gravesite 5484 in Fort Snelling National Cemetery.

Private Bert Hart of Headquarters Company, 132nd Infantry Regiment, who helped carry a wounded Captain Mallon off the field of battle on October 1, 1918, died in 1957. His remains lie in Fort Snelling National Cemetery in Section E, gravesite 718.

PFC Jerome Sauber of Company D, 132nd Infantry Regiment, who was wounded in France, is buried in Section G in gravesite 2430.

CHAPTER NINETEEN
EFFIE AND THE BOYS

Effie Campbell Mallon was left to raise her two boys, George Jr., age 14, and Robert, age 9. On December 34, 1934, just a few months after George Mallon's passing, Bob Greenberg of the Securities Division of the Minnesota Department of Commerce wrote U.S. Congressman Ernest Lundeen on Effie's behalf:

> ... allow me to thank you for anything you have done up to this time, I might say that from my standpoint as a friend of long standing of the Mallon family, I do not see the justice in allowing the widow and the two children some $23.23 per month pension. It is utterly impossible for Mrs. Mallon, who is a woman around fifty years of age, to support herself and the two children on this pension. I cannot for the life of me see why the Federal Government, which allowed Captain Mallon during his lifetime a pension of approximately $165.00, then in turn, after his demise, grant his widow this small sum to support herself and her two children. She has two boys, one if about fifteen years old and the other is around eleven or twelve. The oldest is just entering high school and the youngest is in fourth or fifth grade. Mrs. Mallon has not, during her married life, worked at anything else but as a housewife and to ask her at a time when she is entering the sunset of her life to go out and earn a living for herself and her two boys is something beyond comprehension ... If all the Federal Government can give the widow and two children of one of Pershing's hundred heroes is $22.23 per month, there is something wrong in Denmark.

> Is it not possible for you, as a member of a coming session of Congress, to immediately introduce a bill granting her the same pension rights, namely $165.00 per month for the rest of her natural life, providing she remains unmarried? I know this has been done in several cases. Captain Mallon left no insurance except $500 from the sprinkler fitters union. His government insurance lapsed some two years prior to his death, and the financial circumstances of his wife and two kiddies will bear very rigid investigation from the Department of Justice or whichever department wants to investigate. I certainly think that the United States Government is deeply indebted . . . There are thousands of Captain Mallon's old friends, who are vitally interested in what happens to his widow and two children. In the name of humanity's sake, I trust that you can see your way clear to do something about this very serious matter.
>
> I spoke to Mrs. Mallon yesterday and she told me that she still owes a large grocery bill and will barely have enough money left of the $500 insurance to last perhaps another month or two . . .

U. S. Congressman Ernest Lundeen was well known, not only to his constituents, but to many Americans. His first stint as a U. S. Congressman was from March 1917 to March 1919. He was one of fifty congressmen who voted against the declaration of war against Germany on April 6, 1917. That gained him a certain notoriety.

Effie Mallon wrote Congressman Lundeen on January 19, 1935 to thank him for the wonderful response, which he sent to family friend Bob Greenberg. She added:

> The Gov. has allowed me $23.00 a month for the two boys and myself and you can imagine how far

you can go on that with two boys, nine and fifteen and the way they like to eat. I just don't know what I will do. I had to get help from Welfare (Soldiers) last month and no one knows how that hurt, having been trying to get something to do, but it seems as though there isn't anything for me as I never worked before I was married and, of course, there is nothing much I can do. Everyone has been so kind to me and trying to help me get something and I certainly do appreciate anything that you can do for me. I really feel as my husband has done enough for his country, that the family should be kept off of the Relief. I know from the letter that you wrote Bob, you are doing all you can for us and I want to thank you again.

On March 5, 1935, Congressman Lundeen wrote Effie Mallon:

> We have taken up with the Veterans Administration the matter of your pension and find that the highest rate of pension, which can be paid you under present legislation is $23.00 per month unless it can be shown that Mr. Mallon's death was caused by disabilities contracted in the service in the line of duty.

There was no basis, at the time, for accepting emotional or psychological wounds as trauma.

Minneapolis Labor Review published a column in 1936 about a similar dilemma for Captain Samuel Woodfill, who received the Medal of Honor at Chaumont at the same time that Mallon did. The House Military Affairs Committee, upon recommendation of the Secretary of War, voted down a bill that would have increased Woodfill's retirement pay by $11.25 per month. The Secretary of War commented: "Captain Woodfill never held a commission in the Regular Army. For this reason the War Department objects to enactment of this bill." The *Review* editorialized:

> The whole affair is something to think about on this Memorial Day. And we might also think about the late George Mallon, the Minnesota trade unionist, whose feat of bravery equaled Woodfill's, but who, like Woodfill, was, after the smoke of patriotic fervor cleared away, shoved to the sidelines because he "never held a commission in the Regular Army."

Ernest Lundeen became U. S. Senator from Minnesota on January 3, 1937. He did not forget Effie Mallon's plight. On August 30, 1937, Senator Lundeen wrote Effie Mallon:

> Before Congress adjourned in this last session, we introduced a companion bill to that which was introduced in the House of Representatives by Congressman Teigan granting you a pension. While no action was taken in the Senate before Congress adjourned, we shall follow up this matter when Congress reconvenes for its next session.

Henry Teigan moved from North Dakota to Minneapolis in 1917. He served as the national secretary of the Nonpartisan League through 1923 and came to know George Mallon well. He won a seat in the 75th U.S. Congress in 1937 as a candidate for the Minnesota Farmer Labor Party.

On February 3, 1938, Senator Lundeen wrote Senator George McGill of Kansas, chair of the U. S. Senate Committee on Pensions:

> Reference is made to the special bill I introduced in behalf of Effie G. Mallon, S. 2976, in the 75th

Congress, 1st Session, now pending before your committee. Please let me know what evidence is needed to bring this matter to a favorable conclusion... I wish to assist Mrs. Mallon in every way possible and will appreciate your cooperation.

McGill responded to Senator Lundeen on the following day, February 4, 1938:

... In order that the Bill may receive further consideration, it will be necessary that the enclosed forms be completed and filed with the Senate Committee on Pensions. I may add that no action has yet been taken on bills of this character.

In the end, the pension bill on Effie Mallon's behalf died in committee. She struggled as best she could. By 1940, she was no longer living in the family home on Washburn Avenue just off of Victory Memorial Drive. Effie rented a place at 2700 Penn Avenue N in Minneapolis' Jordan neighborhood. She found work as a filing clerk. Young George was out on his own. Robert was fifteen and living at home. A man was living in the house with Effie and Robert. She married the man in 1945, but the marriage did not last. Effie moved back to Kansas.

The war, in which George Henry Mallon served, did not end all wars. A revival of militarism in Germany led to World War Two in 1939. The United States entered the war in December 1941 to again go to the aid of Britain and France. George Mallon's younger son, Robert Curry Mallon, turned seventeen years old just two months later. When he did enter the service, Robert volunteered for airborne training. He became Private First Class Robert Mallon with Company C, 513th Parachute Infantry Regiment (PIR), 17th Airborne Division. The regiment embarked from Boston on August 20, 1944. It seemed to many that the war in Europe was nearly over. The 17th Airborne Division was still in training in England in September and missed the disastrous Operation Market Garden.

The outlook for the war changed dramatically on December 16, 1944 when Nazi Germany launched a massive surprise offensive in the Ardennes Forest of Belgium and Luxembourg. The 82nd and 101st Airborne were rushed north to stem the flow of German panzers. The 17th Division was flown into France, the last elements arriving on Christmas Day. Patton's Third Army had relieved Bastogne, but the fight was far from over. The German Army's morale remained high and its soldiers put up fierce resistance. Then elite panzer SS grenadiers attacked with tenacity in a desperate attempt to reverse the setback.

PFC Robert Mallon and his comrades in 1st Battalion, 513th Regiment rode north in open trucks for sixteen frigid hours on January 2, 1945. The soldiers were ill equipped for winter warfare. They ate their first hot meal in three days on the evening of January 3, when they received orders to move out at dark. 1st Battalion's objective was the town of Flamierge. PFC Mallon and his comrades in Company C were in the lead on the advance. They stumbled forward single file through nearly two feet of snow in the darkness. The paratroopers' hands and feet were wet and damp and nearly frozen. They had only the ammunition they could carry and they had no offensive antitank weapons. Unknown to the untested men of 513 PIR, they were advancing upon confident veteran German troops supported by tanks.

1st Battalion came under artillery fire as they approached Flamierge in the early morning hours of January 4, 1945. A direct hit on the company commander and his command group cut communication. The company executive officer took command and, shortly afterwards, was wounded. The lead platoons engaged in a fierce fire fight as they moved across 150 yards. After the battalion took the edge of town, they commenced fighting building to building at a heavy cost. The Germans counterattacked with armor and overran the paratroopers. It was not until January 7, with the entire regiment committed, that the 513 PIR securely held Flamierge. The cost was considerable. PFC Robert Curry Mallon was among the many wounded.

After being pulled off the front line, the 513 PIR took part in Operation Varsity, the airborne assault on the Rhine. It was the regiment's first combat jump. 72 C-46s flew the regiment across the Rhine and through the flak to the LZ. The 513th suffered heavy casualties during the drop and in the first moments on the ground. The Germans raked the area with machine gun fire and poured time burst shells onto the position. After heavy fighting, the 513 PIR secured its objective and captured 1,100 prisoners. During this struggle, PFC Mallon was reported missing, then liberated two days later and reported wounded. On April 1, Easter Sunday, the 513 PIR was outside the German town of Munster. The Germans refused to surrender and one last brutal fight ensued before the Americans took the town. The war ended a month later. Robert Curry Mallon retired to New London, Minnesota in 1989. He died on January 14, 2009. He was buried in Fort Snelling National Cemetery in Section B, gravesite 139-2. His remains lie not far from his father's.

By 1954, Effie Mallon was living in the greater Kansas City, Kansas area near extended family. George, the older son, made Minnesota his home, Robert, the younger one, Kansas. Effie died in 1970, thirty-six years after her husband's passing, and was buried in St. Patrick's Cemetery, north of Ogden, Kansas, in the Mallon family plot with her husband's grandparents and parents.

EPILOGUE

On June 2, 1939, five years after his passing, *Minneapolis Labor Review* printed a beautiful tribute to Captain Mallon. The title was "A Man to Remember." The *Review* reminded its readers:

> Minnesota's most famous soldier is the late Captain George H. Mallon ... He was a brave man, but he was a modest man. When time and again he was decorated, he always declared as the decorations were being pinned on him, "You should decorate my men, as well as me. They are just as deserving of decoration as I am."
>
> George Mallon was also a Farmer-Laborite. He made the greatest sacrifice for the Farmer-Labor cause, that perhaps was ever made. When he returned from the war the Republicans offered him the nomination for Governor. The nomination on the Republican ticket at the time assured election. Capt. Mallon chose instead to be candidate for lieutenant governor on the Farmer-Labor ticket ...
>
> [The War] was not his only claim to fame. He delivered the finest addresses in behalf of peace that have been uttered ...
>
> A loyal trade unionist. There is no finer example of a brave and noble man, nor of the unselfishness and sacrifice that built the Farmer-Labor Party than is exemplified in the life of Captain Mallon ... noble and courageous George H. Mallon.

On April 15, 1944, the Farmer-Labor Party and the Democratic Party of Minnesota merged to form the Minnesota

Democratic-Farmer-Labor (DFL) Party. Among the key players was Hubert H. Humphrey, a political science professor at Macalester College in St. Paul. The blending of the "radical" Farmer-Labor Party with the "conservative" Democratic Party created a formidable opposing party to the Republicans in the state of Minnesota.

Bill Thomblison, delegate to the Central Labor Union, wrote a column for *Minneapolis Labor Review* a month later.

> The traditions of Floyd Olson, of Magnus Johnson, of Emil Youngdahl, of Captain Mallon, and of all the others who helped to build a mighty liberal political movement in the State of Minnesota are not dead, despite the shotgun wedding of the Farmer-Labor Party and the Democratic Party . . . they will fling high a common banner in a common cause. And they will be joined by the old timers to build a new party carrying on the same principles that were laid down at the first Farmer-Labor Party convention following the last war.

Minnesota DFL candidate Hubert H. Humphrey became mayor of Minneapolis in 1945, U.S. Senator in 1948 and Vice President of the United States from 1965-1969. DFL Party candidate Orville Freeman, like Mallon a member of VFW and DAV, became governor in 1954. DFLer Walter Mondale served as US Senator from 1964-1976 and Vice President of the United States from 1977-1981.

No simple white marble stone, even one with the inscription, "Medal of Honor," can sufficiently reflect the life of this man. There was so much more to George H. Mallon than Captain Mallon, "One of Pershing's One Hundred Heroes." George Mallon maintained his humility and never let the legend go to his head. Instead, he leveraged the honorifics to advantage in any way that he could to support the causes of veterans, the working class and peace. He lived his life with a purpose. He had his flaws, like any mortal, but he was a noble man.

POSTSCRIPT

By 1947, George H. Mallon had been dead for thirteen years. An interesting article appeared in the Sunday issue of the *Topeka Daily* on September 7, 1947. The title read: "Second Medal of Honor Winner in State Found." A photographer in a previous article credited Medal of Honor recipient George Robb as the only native Kansan to receive the Medal of Honor. A man in Ogden, Kansas, George Mallon's hometown, knew otherwise. He contacted a family member, A. E. Mallon, who delivered news clippings of George Mallon to the Topeka Daily. That realization led to the aforementioned article, which included a photo of Captain George Mallon. The article, which covered Mallon's life and achievements, ended with the statement: "Kansas has at least two men who received the highest of honors in the first World War."

Minnesota, like Kansas, could claim two Medal of Honor recipients. In addition to Mallon, Louis Cukela entered the military service from Minneapolis. Cukela received his award for heroism on July 18, 1918 near Soissons while a sergeant with the 66th Company, 5th Regiment Marines.

Both Kansas and Minnesota have good ground on which to "claim" Captain George H. Mallon as one of theirs. George was born and raised just outside Ogden. He met and married Effie Campbell in Kansas. His parents and grandparents are buried near the family place, as is George's wife Effie. Mallon first served in the Kansas Volunteers during the Spanish American War. At the same time, Mallon lived most of his adult life in Minneapolis. He went to World War One from Minnesota and returned home to Minnesota. However, his wife Effie lived in Kansas City during the war. He was active in politics and in labor in Minnesota for the post-war years of his life. However, he did spend a year in Kansas, helping to organize the Nonpartisan

League among Kansas farmers. In the end, George Mallon died and was buried in Minnesota. He belongs to both Kansas and Minnesota, but also to the United States of America in whose army he served as a commissioned officer.

George Mallon's wife Effie and her two sons were in possession of Captain Mallon's actual medals and uniform and other personal effects. The family donated these to the Military Museum at Fort Riley, Kansas. The Minnesota Military History Museum at Camp Ripley, near Brainerd, Minnesota, is the repository for Captain George Mallon's award documents. Professionals restored and conserved the documents.

SOURCES AND ACKNOWLEDGMENTS

The primary source of information and photos for this book was the Mallon Family, who willingly shared with me all that they had. This book came about because of their wonderful response. These are the descendants of George and Robert Mallon, the two sons of Captain George H. Mallon and his wife Effie Campbell Mallon, and of George and Effie's siblings, including George Mallon, Diane Gossage, Rod Rolfe, Kay Mallon Harris and Jennifer Meuhlbach. Thanks to Hennepin County Library, an extraordinary institution, and the frequent efforts of the staff of InterLibrary Loan and also of Ted Hathaway and his staff in Special Collections. Thanks also to the reference staff of the library and research center at the History Center of Minnesota Historical Society. Thanks to Major Doug Bekke (Retired, USA) and Steve Osman, retired site manager of Historic Fort Snelling, for sharing their insight on World War I, including the Meuse-Argonne battlefield at Forges Creek. I very much appreciate the steadfast support of my writing and speaking by Colonel Don Patton (Retired, USA), founder and head of the Dr. Harold C. Deutsch World War Two History Roundtable and the support of Lt. Colonel John Knapp (Retired, USMC), Director, Fort Snelling National Cemetery. This book would never have been possible without the infinite patience and unfailing encouragement of my loving wife, Mary Ann. I also want to acknowledge my late father, Duane, for his willingness to drive off the beaten path on family vacations to allow me to visit so many historical sites, which were obscure to anyone but a passionate student of American history.

CHAPTER ONE: IRELAND
The Mallon Family genealogy provided the basis of this chapter.
Some books on the history of Ulster and the problems between Catholics & Protestants include:
Beckett, J.C., *The Making of Modern Ireland, 1603-1923* (published in 1966),
Donnelly, James S., *The Great Irish Potato Famine* (2001)
Elliott, Marianne, *The Catholics of Ulster: A History* (2001)

Stephen D. Chicoine

CHAPTER TWO: KANSAS

The efforts of the following are appreciated: Linda Glasgow, Curator of Archives & Library, Riley County Historical Museum, Manhattan, Kansas; Joel A. Meyers, Independent Family Historian & Genealogist, Manhattan, Kansas; Karen Bonar, Editor of *The Register*, Catholic Diocese of Salina, Kansas and the Kansas State Historical Society.

The photos of Captain Mallon's medals are courtesy of Deborah J. Clarke, Museum Specialist, and Robert Smith, Museum director and supervisory curator, Fort Riley, Kansas.

Sources for the history of early Central Kansas include:

Thomas, Sister M. Evangeline, "The Rev. Louis Dumortier, S. J., Itinerant Missionary to Central Kansas, 1859-1867," *The Kansas Historical Quarterly*, v. 20, November 1952, pages 253-270.

Jeffries, John B., *An Early History of Junction City, Kansas*, M.A. thesis, Kansas State University, 1963.

Thomas, Delores and Weller, Doug, "Origins, Ogden Parish, the oldest in the diocese, to celebrate 150 years," *The Register*, Catholic Diocese of Salina, February 27, 2009.

The military service of George Mallon's grandfather George Stephens came from Mallon descendant Jennifer Muehlbach.

Of the many articles on Conrad Schmidt, Medal of Honor recipient and Fort Riley range rider, among the best are: *Junction City Weekly Union*, October 1, 1881, page 5, *Junction City Weekly Union*, July 9, 1894, page 4, *Junction City Weekly Union*, January 15, 1897, page 1, a detailed account of his Civil War heroics in *Junction City Weekly Union*, Sept. 8, 1905, page 3, and Conrad Schmidt's obituary on page 1 of the January 1, 1909 issue of *Junction City Weekly Union*.

CHAPTER THREE: SPAN AM WAR & THE PHILIPPINES

The excerpts from the letter, "Christmas in Luzon", written by Cecil Taylor, 12th U.S. Infantry, December 24, 1899, was accessed from the New Bern-Craven County (North Carolina) Public Library.

http://newbern.cpclib.org/research/spamwar/taylor_18991224.htm

The story of the baseball games between the 12th U.S. Infantry and the 25th U.S. Infantry is from John Henry Nankivell's classic *Buffalo Soldier Regiment: History of the Twenty-Fifth United States Infantry* (1927). Nankivell served as captain, commanding the regiment.

Gerald Gems, author of *Sport and the American Occupation of the Philippines* (2016), offered insight.

The newspaper reference to Mallon's belt from the boxing championship in the Philippines is from page 7 of the April 14, 1920 issue of *Willmar Tribune*.

The Theodore Roosevelt quote is from page 1, column 7 of the November 22, 1898 issue of *New York Journal and Advertiser*.

CHAPTER FOUR: FLURRY OF PUNCHES

The most complete source for the 1904 St. Louis World's Fair is the massive 800-page volume by Mark Bennitt and Frank Stockbridge, *History of the Louisiana Purchase Exposition*, published in 1905.

The article on Jack Dunleavy and hooks is from page 26 of the Sunday, December 15, 1901 issue of *The St. Louis Republic*.

Captain Mallon Doughboy Hero

The article on George Mallon's early success in amateur boxing, including the sketch of Mallon's uppercut punch, is from "Mallon Exponent of Popular Fighter in Amateur Boxers' Class" page 2 of the March 3, 1905 issue of the *St. Louis Post-Dispatch*.
One of the best articles on the Mallon-Kennedy fight is "Punishment No Bar" from the page 1 of the July 5, 1906 issue of The Kansas City (Kansas) Globe. Another good article is "Mallon Was Knocked Out" from page 3 of the July 6, 1906 issue of *The Junction City (Kansas) Weekly Union*. The article on George Mallon wanting a rematch with Spike Kennedy is from page 1 of the July 26, 1906 issue of *The Leavenworth Weekly Times*.

CHAPTER FIVE: MINNEAPOLIS TRADESMAN

Minneapolis Labor Review was an important source for Mallon in his pre-war work. Notice of Mallon being elected delegate for the Sprinkler Fitters to the Business Trades Council was on page 1, column 1 of the May 9, 1913 issue of *Minneapolis Labor Review*. The article on the Fat Man's Race is from page 1, column 5 of the August 27, 1915 issue of *Minneapolis Labor Review*.
The article about Brother Mallon going to war is from page 20 of *Plumbers, Gas and Steam Fitters' Journal*, volume 21, no. 6, June 1917, published in Chicago. A copy of the page is part of the Mallon Family Papers. On page 62 of the same issue, the Journal lists Mallon as Business Trade Agent.
The final quote of the chapter, Mallon telling his fellow union members that he hopes he will be back to work for Labor and Van Lear in the next election, is from page 3, column 1 of the August 24, 1917 issue of *Minneapolis Labor Review*. The same piece cited Mallon as "the hustler of hustlers in the Van Lear campaign."

CHAPTER SIX: GOING OVER THERE

The accidental death of Captain Hogstedt at Sour Lake is covered in the *Houston Daily Post* of November 9, 1917 on page 7, column 3.
For more information on the Oilfield Strike of 1917, see *Handbook of Texas Online*, James C. Maroney, "Oilfield Strike of 1917," http://www.tshaonline.org/handbook/online/articles/doott. Also see, Laurie, Clayton and Cole, Ronald, *The Role of Federal Military Forces in Domestic Disorders, 1877-1945*, CMH Pub. 30-15, U.S. Army Center of Military History, 1997, pages 250-252.
The activities of the Prairie Division in Texas, including Captain Mallon's assignment to command Company E, 132nd Regiment is from the *Houston Daily Post*, January 5, 1918, page 7, column 1.
The article from Kansas City about Mallon working hard to improve Company E is from an unmarked, undated newspaper clipping in the Mallon Family Papers.

CHAPTER SEVEN: HAMEL

Bean, C. E. W., Volume VI – The Australian Imperial Force in France during the Allied Offensive, 1918 (1942), Volume VI – Official History of Australia in the War of 1914-1918, pp. 242-335.
Huidekoper, Frederick L., *The History of the 33rd Division, A.E.F.* (1921), Illinois State Historical Society, pp. 115-117. "Report of Captain George H. Mallon, 132nd Infantry, on operations of July 4-5, 1918 at Hamel and Vaire"
Nunan, Peter, "Diggers' Fourth of July", *Military History* (2000), vol. 17, issue 3, pp. 26-32, 80.

CHAPTER EIGHT: MEUSE-ARGONNE
George H. Mallon's four-page post-war State of Minnesota Military Service Record through the Minnesota Historical Society was an important source of confirmation of dates.
The most important source on Mallon's regiment & division in the Meuse-Argonne Offensive is:
Huidekoper, Frederick L., *Illinois in the World War: The History of the 33rd Division, A.E.F.*, Illinois State Historical Society, 1921. Huidekoper was lieutenant colonel and division adjutant.
Volume I/Chapter III, Beginning of the Meuse-Argonne Battle, pages 76-83.
Volume I/Chapter III, Defense of the Meuse Sector, pages 84-91.
Volume III/Appendix XXII, Chronology & Operations for 132nd Regiment, pages 145-163.
An additional source was a publication of the 132nd Regiment, covering its operations in France in 1918 and listing those killed or wounded, from the Mallon Family papers. *Minneapolis Labor Review* gave George Mallon considerable coverage during the war. The online archives of *Minneapolis Labor Review* were a very useful resource in this regard. Mallon was not even on the radar screen of the other Twin Cities newspapers until he was a recipient of the Medal of Honor. The *Review* reported on Mallon's heading off for war and on his officer's commission in "Captain Mallon" on page 3, column 1 of its August 24, 1917 issue. The *Review* published a letter, which Mallon sent, on his being wounded on page 1, column 4 in its November 15, 1918 issue. The *Review* published excerpts of a letter from Mallon announcing his being released from the hospital and heading off to Germany to re-join his regiment in the occupation on page 4, column 1 of the February 28, 1919 issue.
The names of the men, who were with Captain Mallon when they captured the four big howitzers is from an undated newspaper clipping from the Mallon family. The article is from a November 19, 1918 cable sent by Junius B. Wood, a foreign correspondent for the *Chicago Daily News*, who covered the A.E.F in France in 1918. Wood later covered the Soviet Union in the 1920s, the Japanese in Manchuria in 1932 and personally witnessed the rise of Hitler in 1933-1934.
"Hero Chaplain of the World War Died Suddenly, obituary for Father John L. O'Donnell, *Catholic Courier and Journal*, May 17, 1929, page 1, column 5.

CHAPTER NINE: MEDAL OF HONOR & ONE OF PERSHING'S 100
The mainstream press provided considerable coverage for the Victory Labor Loan and Pershing's One Hundred Heroes.
For more on Liberty Loans, see the Federal Reserve History website article titled, "Liberty Bonds, April 1917 – Sept 1918 https://www.federalreservehistory.org/essays/liberty_bonds.
An extensive article on the Liberty Loan Publicity Campaign can be found at http://www.theodora.com/encyclopedia/l2/liberty_loan_publicity_campaigns.html.
The photo of Captain Mallon and the other recipients in line at Chaumont to receive the Medal of Honor from General Pershing is from page 512, *American Armies and Battlefields in Europe*, prepared by the American Battle Monuments Commission and published in 1938. The exploits of Mallon and Gumpertz, which led to their decoration, are on page 199 of the same volume.

CHAPTER TEN: WORLD WAR VETERANS

References in newspapers of the day to the World War Veterans are relatively uncommon. In contrast, *Minneapolis Labor Review* gave the World War Veterans good coverage. The online archives of *Minneapolis Labor Review* were a very useful resource in this regard.

The article about George Mallon becoming Business Agent for the Minneapolis Building Trades Council is from page 1, column 1 of the July 18, 1919 issue of *Minneapolis Labor Review.*

The article on Mallon's noon address at Minneapolis Steel & Machinery Company was on page 1 in columns 3, 4 & 5 of the September 19, 1919 issue of *Minneapolis Labor Review.*

The reply of Harrison Fuller, state commander of the American Legion, to the challenge to debate by Lester Barlow of the World War Veterans appeared on page 8, column 5 of the October 3, 1919 issue of *Minneapolis Morning Tribune.*

The pamphlet **The Kind of Men Behind the World War Veterans** is from the Mallon Family collection.

The article on Captain Mallon's position on universal military training was on page 1, column 1 of the November 14, 1919 issue of *Minneapolis Labor Review.*

The uproar over the raid on the IWW office and the mob assault on Ernest Lundeen can be found throughout page one of the November 21, 1919 issue of *Minneapolis Labor Review* and continuing to other pages.

The welcoming reception of de Valera is from page 1, column 2 (continued on page 2) of the October 20, 1919 issue of **The** *Minneapolis Morning Tribune.* The article of de Valera's address in Minneapolis on October 20. 1919 is from columns 2 & 3, page 4 of the October 21, 1919 issue of **The** *Minneapolis Morning Tribune.*

CHAPTER ELEVEN: NONPARTISAN LEAGUE POLITICS

The *Nonpartisan Leader* article on the refusal of the main Twin Cities newspapers to accept Mallon ads appeared on page 4 of the July 5, 1920 issue. **The** *Nonpartisan Leader,* the League's newspaper, published in Fargo, North Dakota from 1915-1921, is an important primary source.

The "Reign of Terror Condemned by War Veterans" front-page headline was in the January 23, 1920 issue of *Minneapolis Labor Review.* The article on Mallon's address demanding free speech began on page 1, column 6, continued on page 4, column 1.

There are several books on the Nonpartisan League. The finest of these is **Insurgent Democracy: The Nonpartisan League in North American Politics** by Michael J. Lansing (2015). An older book of value is **Political Prairie Fire: The Nonpartisan League, 1915-1922** by Robert L. Morlan (1955). Carol Jensen published an important article in **Minnesota History**, Summer 1972, titled, "Loyalty as a Political Weapon: The 1918 Campaign in Minnesota." Another is "Things as They Should Be: Jeffersonian Idealism and Rural Rebellion in Minnesota and North Dakota, 1910-1920" by Larry Remele in **Minnesota History**, Spring 1988.

George Mallon's appearance in New Ulm on May 23 was featured in a lengthy story in the May 26, 1920 issue of **New Ulm Review**, beginning on page 1, column 7 and continuing on columns 1 through 4 on page 2.

A number of Minnesota newspapers outside of the "kept press" in the Twin Cities provided the move by the Nonpartisan League to file to run as a third party after losing the Republican Primary in June. These included *Labor World* (Duluth, MN), the ***Princeton Union*** and the ***Willmar Tribune***.

The *Minneapolis Labor Review* editorial, which highlighted George Mallon as having "the hardest of any of the ticket" in what was "practically a man to man fight between Mallon and Collins" appeared under the headline, "A Real Lieutenant Governor" on the top of column 1, page 3 of the paper's June 11, 1920 issue.

Discussion of the formation of the WPNPL as a counterpart of the NPL and that it would not have been allowed in 1918 is from "Labor and Farmers in Minnesota Join Hands Politically" by A. Karlson on pages 12-13 in the May 1, 1920 issue of *Brotherhood of Locomotive Firemen & Enginemen's Magazine*.

The letter written by plumber Bob Morgan in support for George Mallon on the eve of the primary appeared on column 5, page 7 of the June 19, 1920 issue of *Labor World* (Duluth, MN).

CHAPTER TWELVE: RETURN TO KANSAS

Several Kansas newspapers were a source of the story of Mallon in Kansas.

Major newspapers as prominent as and from as far away as The *Washington Post* reported on the Saline County Legion Post and Captain Mallon with the Nonpartisan League. See page 2 of the January 11, 1921 issue of The *Washington Post*.

The reference to Mallon carrying with him his grandfather's GAR button is from the top of columns 5 & 6, page 2 of the January 27, 1921 issue of *The Kansas Leader* (Salina, Kansas).

Two articles, "To Question Them" and "Refused to Let Them Speak" on columns 3 & 4 of page 12 of the March 17, 1921 issue of The *Topeka State Journal* covered the conflict in Marion County, Kansas in which Mallon became involved when a mob refused to let Burton and Wilson speak in Barton, Kansas. *The Kansas Leader* of the same date also gave the incident a full page of coverage on page 1.

One of the best and most comprehensive newspaper articles concerning the Nonpartisan League is "Getting the Truth About the League" on page 6 (the entire page) of the March 21, 1921 issue of *The Nonpartisan Leader* (Fargo, N.D.) and continuing of page 7 and 17.

The move of Mallon from radical organizing to the insurance business is from page 2, column 5 of the August 26, 1921 issue of *Minneapolis Labor Review*. Similar announcements appeared in smaller newspapers across the state, including *Labor World* (Duluth) on September 3, 1921, page 17.

The article on Mallon speaking at the 1921 Labor Day event is from page 1, column 1 of the September 9, 1921 issue of *Minneapolis Labor Review*.

"Landslide Didn't Touch Nonpartisan League" in *The Kansas Leader* (Salina, KS) of December 9, 1920 on page 5 made very clear that the Nonpartisan League gained a considerable number of seats in the Minnesota legislature, despite the Republican landslide, which swept Harding into the White House.

CHAPTER THIRTEEN: VICTORY MEMORIAL DRIVE

Minneapolis Morning Tribune of June 12, 1921 covered the dedication and the August 11, 1921 issue covered Pershing's visit.

The newspaper photo of George Mallon and Louis Collins is from page 1 of the June 16, 1921 issue of American Legion newspaper *Hennepin County Legionnaire*, courtesy of Al Zdon of the American Legion Department of Minnesota in St. Paul.

CHAPTER FOURTEEN: TOMB OF THE UNKNOWN SOLDIER
VFW's *Foreign Service Magazine* of December 1921 proudly announced the presence of Captain George Mallon at the dedication of the Tomb of the Unknown Soldier on page 11.

The ACLU press release about the Joint Amnesty Committee is from "Soldiers' Organization Takes Action Urging Debs' Release," *Brotherhood of Locomotive Firemen and Enginemen's Magazine*, vol. 71, no. 12, December 15, 1921, p. 3.

CHAPTER FIFTEEN: COUNTY COMMISSIONER
Kris Leinicke of the Rock Island Arsenal Museum graciously researched the disposition of the German anti-aircraft gun, which George Mallon captured on September 26, 1918, and provided documents to support the scrapping of the gun in 1942.

The election of officers for the Minnesota VFW is from an article on column 1 of page 20 in the August 11, 1921 issue of the *Minneapolis Morning Tribune*.

Mallon is mentioned regarding the Mendota Bridge as early as 1923. For example, page 2, column 1 of the September 21, 1923 issue of *Minneapolis Labor Review*.

CHAPTER SIXTEEN: HERO OF A DISTANT PAST
For more information on Fawcett Publications, see Minnesota Historical Society's Mnopedia article
http://www.mnopedia.org/person/fawcett-wilford-hamilton-captain-billy-1885-1940.

Also see the Wikipedia article on Fawcett https://en.wikipedia.org/wiki/Fawcett_Publications.

The story on Captain Mallon in the April 1931 issue of **Battle Stories** (vol. 8, #44, pp. 126-132) was sourced from library of the Minnesota Historical Society. The collection call number for Fawcett Publications' monthly publication **Battle Stories** is AP2 F27. An excellent index to the Battle Stories issues can be found online at http://www.philsp.com/homeville/afi/b6.htm#A71.

The discs containing the Rickenbacker interviews for the Chevrolet Chronicles seem to no longer exist or are in private collections. The transcript from Rickenbacker's interview with Mallon was obtained from the Mallon Family.

NEA Service was a newspaper syndication service delivering content to some 700 newspapers in the early 1930s. The nearly full-page feature article, "Where are General Pershing's Ten Greatest World War Heroes Today?" appeared in smaller town newspapers across the nation on November 11, 1931. Among these were *The Brainerd Daily Dispatch* (Minnesota), *Manitowoc Times* (Wisconsin) and *The Brownsville Herald* (Texas). Thanks to David Parsons of Brownsville Historical Association for providing me with a really copy of the article and of the photo of Mallon and his two sons.

The Loring School article and photo are from the Mallon Family collection.

CHAPTER SEVENTEEN: THE FINAL MONTHS
George Mallon's health history was sourced from his death certificate, sourced from the Minnesota Historical Society. Very few death records include such detail. This must have been the result of the extended visits, which Mallon had to the Veterans Hospital. Sinclair Lewis' visit to the Veterans Hospital in St. Cloud is from the St. Cloud Daily Times, Saturday, July 28, 1934, page 3, column 2.

Stephen D. Chicoine

The finest tribute to George Mallon's life can be found in *Minneapolis Labor Review*, August 10, 1934, page 4, column 1 under the headline, "A Brave Man Dies." *Minneapolis Journal*, August 3, 1934, page 1, column 3, continued page 6, published a lengthy obituary under "George Mallon, One of U.S. 100 War Heroes, Dies." *St. Cloud Daily Times* included a notice on page 3, column 2 of the Friday, August 3, 1934 issue.

The story of the Bonus March has been told many times. Jennifer Keene did a fine job of placing the march in context in her 2003 book *Doughboys, the Great War and the Remaking of America*. A great resource is Paul Dickson and Thomas Allen's 2004 book *The Bonus Army: An American Epic*. The online archives of *Minneapolis Labor Review* are a great source of information on the Bonus March

The story of Thomas Van Lear's death and funeral is from the online archives of *Minneapolis Labor Review*. The article was on page 1, column 1 of the Friday, Marsh 13, 1931 issue.

The article about Ben Neff as treasurer for the DAV in 1954 was on page 6, columns 5 & 6, of the December 9, 1954 issue of *Minneapolis Labor Review* in an article titled, "DAV To Open New Home on Friday."

The information on the priests, who served as celebrants at Mallon's 1934 funeral and burial at St. Mary's Cemetery is from Sehri Strom, Records Manager, Office of Archives and Records Management, Archdiocese of Saint Paul and Minneapolis.

CHAPTER EIGHTEEN: FORT SNELLING NATIONAL CEMETERY

The correspondence by and on behalf of Theresa Ericksen is courtesy of John Knapp, Director of Fort Snelling National Cemetery. The information on Theresa Ericksen is from a host of sources, including Jean Shulman, R.N., Volunteer Nurse Historian, American Red Cross, Vickie Wendel of the Anoka County Historical Society and the files of the Minnesota Historical Society.

CHAPTER NINETEEN: EFFIE & THE BOYS

The correspondence concerning Effie Mallon's pension is from the Ernest Lundeen Collection, Box 367. Folder 1, courtesy of Carol Leadenham, Assistant Archivist for Reference, Hoover Institution, Stanford University.

CHAPTER TWENTY: EPILOGUE

The article on two Medal of Honor recipients from Kansas, "Second Medal of Honor Winner in State Found," was in *Topeka Daily* on September 7, 1947.

Stephen D. Chicoine is the author of many books, including *Our Hallowed Ground: The World War Two Veterans of Fort Snelling National Cemetery*, *The Confederates of Chappell Hill, Texas* and *John Basil Turchin And the Fight to Free the Slaves*. He is a graduate of the University of Illinois and Stanford University. Mr. Chicoine speaks regularly at a wide range of venues. His forthcoming series, *Glory, Tragedy & Trauma: Stories from A National Cemetery* will include volumes on Bluecoats & Buffalo Soldiers, Doughboys, World War II, Korea and Vietnam.

Mr. Chicoine's author website is freedomhistory.com.

www.ingramcontent.com/pod-product-compliance
Lightning Source LLC
Chambersburg PA
CBHW070601300426
44113CB00010B/1347